The Yankees' Secret Weapon

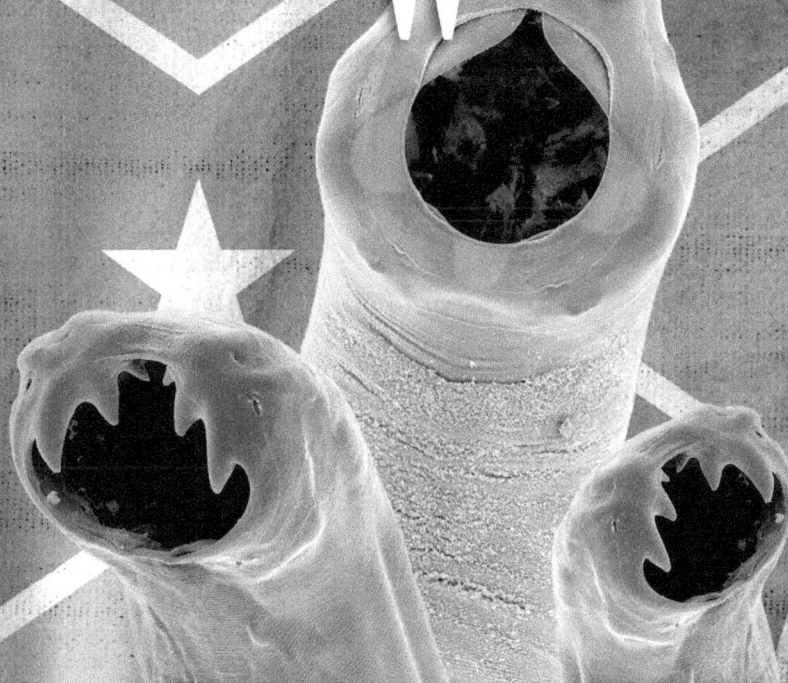

EVEN LINCOLN DIDN'T KNOW

Thomas Power Lowry M.D.

HISTORY BOOKS BY THE AUTHOR

The Story the Soldiers Wouldn't Tell

The Attack on Taranto

Tarnished Eagles

Civil War Bawdy Houses

Lincoln and Military Justice

Tarnished Scalpels (with Jack Welsh)

Swamp Doctor

VD and the Lewis & Clark Expedition

Confederate Heroines

Sexual Misbehavior in the Civil War

Mystery of the Bones (with P Willey)

Andersonville to Tahiti

Chinese Soldier in the Indian Wars

Confederate Death Sentences (with Lewis Laska)

Love & Lust: Intimate Civil War Letters

Merciful Lincoln

Utterly Useless Union Officers

Bad Doctors (with Terry Reimer)

Irish & German—Whiskey & Beer

Capital Courtesans

Civil War Rockets

Titanic Madness—Alzheimer's Caused?

Lost Lincolns

Civil War VD Hospitals

Thousand Stories You Don't Know

More Stories You Don't Know

THE YANKEES' SECRET WEAPON

EVEN LINCOLN DIDN'T KNOW

Thomas Power Lowry, MD

Idle Winter Press
Portland, Oregon

Idle Winter Press
Portland, Oregon
http://IdleWinter.com

This edition published 2016
Printed in the United States of America
The text of this book is in Alegreya

ISBN-13: 978-0692682449 (Idle Winter Press)
ISBN-10: 0692682449

CONTENTS

ACKNOWLEDGMENTS

Robert K. Krick, in this and many other books, is my source for all things Confederate. The Rev. Dr. Albert H. Ledoux is a master of family and geological research. The Rockefeller Archives were central in this work. Beverly A. Lowry, co-researcher and long-time companion, saved me from a host of errors with her proofreading.

INTRODUCTION

John D. Rockefeller, Sr. knew about it. Thomas Edison knew about it. US Army surgeons knew about it. The governor of North Carolina was told about it, but didn't believe it. This secret weapon was totally unknown to Lincoln. Neither he nor his Secret Service chief, Allan Pinkerton, knew about it. Even Lincoln's chief spy in Richmond, Elizabeth van Lew, was unaware of it.

The great minds of 1861-1865 may be forgiven—this phantom menace was not fully understood until around 1902. Its most vivid manifestation was seen in the 1950s by US Army MASH surgeons in Korea. The American infantry ranks were filled out by KATUSAS—Korean Augmentation to United States Forces. When these Korean comrades were gut-shot and brought to the operating table, the instrument tray always contained an empty and sterilized No. 10 can. As soon as the doctor opened the belly, he dipped his hand into the opening and began to scoop out handfuls of worms. Long worms. Short worms. White worms. Pink worms. Slimy wiggling worms, tumbling over each other in their blind panic. Only after all the worms were gone, dumped into the empty can, did the surgical team begin the repair of the soldier's torn intestines.

The Koreans, like most Third World people, were infested with intestinal worms: round worms, pin worms, tape worms, flat worms, and—most germane to our study—hookworms. The hookworms were the Secret Weapon of the Civil War, one that weakened the soldiers of the Confederacy far more than the men in blue, for reasons soon to be made clear.

And why does John D. Rockefeller fit into our tale? Today he is most remembered for his rapacious creation of consolidations, monopolies, cartels, and trusts, his building of an oil empire, which lives on today, in re-named fragments such as Chevron, leading to the largest accumulation of wealth in American history. Yet this deeply religious man was equally dedicated to another, less well-remembered career, the formation and funding of a great diversity of charities, including Spelman College, Yale, Harvard, The University of Chicago, and the Rockefeller Institute for Medical Research. The charity most relevant to our thesis was the Rockefeller Sanitary Commission, which functioned from 1909 to 1915, and focused on hookworm epidemiology and treatment throughout the Deep South. The findings of this commission can tell us much about the fighting efficiency of the boys in butternut and gray.

In the chapters to follow we will take up sixteen different questions about hookworm disease in the Civil War: What is this *Necator* beast? How big is it? How does it get into the human intestine? How does it harm its host? Is it the biology behind the term "poor white trash?" Why was it more prevalent in the South? Why did the presence of "Negro blood," as Southern legislatures termed persons of African origin, protect the bearer? Further questions included: why were the design and function of "sanitary latrines" so important? How did such facilities impact white schools when compared with "Negro" schools? And finally, can the data of the 20th century be reasonably extrapolated to the health and vigor of the soldiers who defended the Confederacy through four bloody years of combat?

Our current movies of the Civil War are full of Confederate re-enactors who appear pink-cheeked, lively, and well-nourished. Is this an accurate portrayal of those long-ago men of the South? Let us see where the facts lead us.

CHAPTER 1

MR. NEMATODE'S WILD RIDE[1]

Much of life is earthy, crude, raw, and often ugly. The *Necator americanus* animal begins its cycle in the lowliest of spots, in the earth itself, and mixed with what the Spanish call *la mierda*. In French we see *la merde,* and in German *Scheiße* The ancient Greeks called it κόπρος. In polite conversation, it might be "fecal matter," but no matter what the name it's all the same stuff, the essential world for the hookworm. Into this substrate of warm, moist Southern soil comes the egg. Like the foolish riddle of "What comes first, the chicken or the egg?" the answer is "a circle." Chickens beget eggs and eggs beget chickens. The life of a hookworm, too, is a circle. A complex circle.

It begins with John Greenleaf Whittier's "Barefoot boy with cheek of tan." This shoeless lad, or any other unshod human, places his naked foot on to a patch of earth which has been soiled with *dreck* (If one summons Yiddish to our thesaurus). If an egg (or several thousand eggs) were planted there several days prior, the hapless pedestrian encounters the earth-bound forms of *Necator*. The eggs have hatched into rhabditiform larvae, which after two molts (or ten days) have evolved into filariform larvae. This second larval stage has the capacity to penetrate the skin. They can penetrate unbroken skin, but prefer entering through a hair follicle, in both cases using a proteolytic enzyme, literally digesting their way into the closest blood vessel. And they do so, leading to the next steps, which rival any Halloween horror movie.

These tiny thread-like worms enter the venous capillaries of the foot and continue upward into the larger veins, which join the iliac veins, which in turn join to form the massive inferior vena cava. This ascending torrent of blood pours into the heart, bearing its dreadful load of tiny wiggling larvae. It encounters first the right auricle. With each heartbeat the auricle's contents are ejected into the right ventricle. This powerful muscular chamber, with the next beat, forces blood into the lungs, via the pulmonary artery, which branches into smaller and smaller arteries and finally into tiny capillaries, which surrounded the alveoli. These microscopic air sacs are where inspired air conveys its life-giving oxygen into the blood. The blood of the pulmonary artery, as it passes through millions of pulmonary capillaries, gives off carbon dioxide which is exhaled with the next breath and picks up oxygen. The now-oxygenated blood collects in the pulmonary vein and enters the left side of the heart, where it is pumped into the aorta, on its way to nourish the entire body.

Unfortunately for the barefoot lad (or his father) who put his foot in the wrong place, the pulmonary capillaries give up more than carbon dioxide. The agile larvae, when they reach the smallest capillaries, are larger than the capillary itself. When they "feel" this tight fit, they release another proteolytic enzyme and dissolve their way into the adjacent air sac. From there they climb up and up, first into the small bronchioles, then the large bronchi, and finally, the trachea. Anyone who has ever accidentally inhaled a speck of dust, and experienced the violent coughing that even a mote may call forth, can sympathize with the Confederate soldier who feels these agile, wiggling little invaders grabbing the lining of his windpipe, hauling themselves upwards against his attempts to expel them. (One text uses the euphemistic words "pulmonary symptoms.") Finally, they reach their Mount Everest—the larynx. Once there, they slither between the vocal cords, cross the small space that separates the cords and the esophagus and are swallowed down into the stomach. The *Necator* are astonishingly adaptable. It's no surprise that blood and tracheal linings might be a pleasant environment for them, but stomach acid, a much different chemical mix, fazes them not one whit. They merrily pass this caustic gauntlet and enter the upper small intestine where they find a happy home, using their tiny cutting plates (seen in the cover photo) to chomp into the intestinal wall.

The lining of the small intestine is no flat surface; rather it contains millions upon millions of *villi*. Each of these protrudes into the cavity of the gut and resembles a microscopic finger. Tiny capillaries and lymphatic channels within the villous absorb nutrients from the food that the host has eaten. Now, the host himself gets eaten. To the hookworm this waving meadow of villi is an all-you-can-eat buffet. It takes a single worm about four days to devour a single villous, deriving nutrition by digesting not only the villous tissue but also the blood released by the severed capillary. When he (or she) finishes dining on that one villous he moves on to the next.

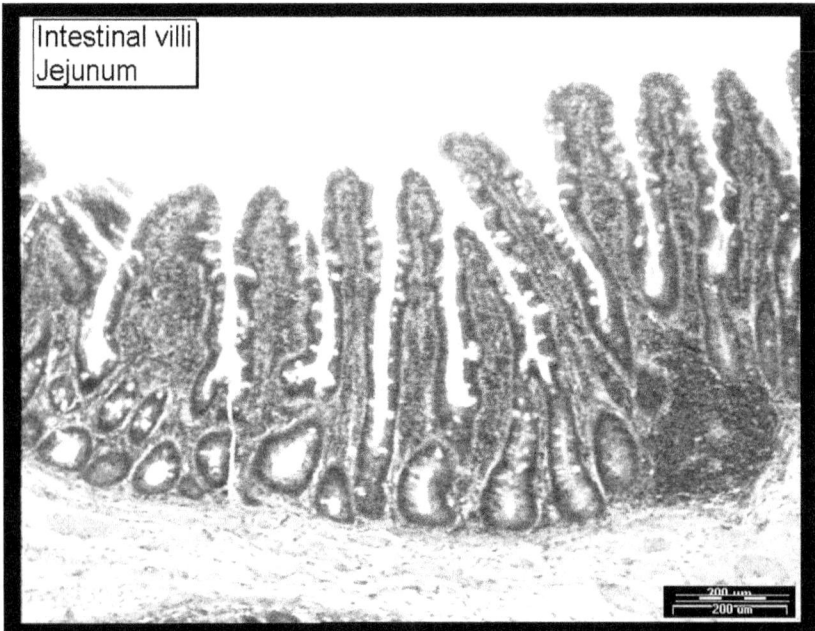

Intestinal villi
Jejunum

200 um

The mention of gender is important. As they attach to the gut lining and begin gnawing their way to glory, they pair up, one male, one female, connect their mutual reproductive organs and set to making eggs. This conjugal embrace lasts their lifetime. Not even Tantric yoga can yield such duration of attachment. And the fruit of this union? They more than match the admonition in Genesis 1:22. The happy couple, in their years together, will produce roughly 200 million eggs. A typical infestee might carry 2,000 hookworms. Two thousand times 200 million adds up to a whole lot of hookworm eggs. Most soldiers, both officers and men, had little regard for today's notion of sanitation. Men relieved themselves wherever convenient. Those with diarrhea, bedeviled by cramps and urgency, never made it past their tent door. Nostalgic re-enactors, singing "Tenting Tonight on the Old Campground," are blissfully unaware of the horrendous mess that a regiment of a thousand men could, and did, make.

The song itself, re-created beautifully on YouTube, speaks volumes of sadness and loss, and only the most cold-hearted remain unmoved, but parasites are not noted for sentimentality and so we return to the world of biology. Each adult worm averages only ten millimeters in length and removes only a small drop of blood every day, but thousands of drops of blood every day, month after month, year after year, add up to anemia. The pale, weakened victim simply does not have enough blood, enough red cells, to carry the vital oxygen his body needs. The victim looks pale because he is pale. He feels weak because he is weak. The Civil War soldier was assaulted by the well-documented plagues of malaria, measles, smallpox, dysentery, mumps, and syphilis. His wounds often turned gangrenous, the first step on the path to the cemetery. But what was the role of the hookworm, unknown during the war itself, in weakening the Confederate soldier?

What is needed next is some reliable estimate as to how many Southern soldiers suffered from this unseen enemy, an estimate derived from the knowledge of the following century Just as the war had its famous generals, the war on hookworms had its own Great Men.

---

1. With apologies to Walt Disney and Kenneth Grahame.

CHAPTER 2
HOOKWORM HEROES

The general public tends to have inaccurate ideas of science, inflamed by headlines of "breakthrough," and "miracle drug," or through television soundbites about Noble Prize winners and their cash awards. Other "information" comes from political candidates who fancy themselves experts on biology. Average citizens' eyes glaze over at news of the large hadron colliders, whose findings can be expressed only in equations. Discoveries seem sudden, because the preceding years of patient work attract no reporters. Science could be better compared with the experience of a person on a small boat in a dense fog. He (or she!) senses that something is "out there." Slowly that something emerges from the fog. At first the object is only a dim shape, a change of pattern in the swirling mist, but then, bit by bit, the object can be seen more clearly, and at last it stands out, sharp and clear.

The world of hookworm follows that pattern. Bits and pieces of understanding emerged, which encouraged others to build on that early knowledge. First findings may need correction or revision. Science is not a *thing*; it is a process. Good science rests on evidence and credits those who came earlier. This is not to say that scientists are immune to rivalry, stubbornness, and other sins of human endeavor. History is full of "I discovered it first," which in previous centuries resulted in bitter feuds. Today, legions of patent attorneys join the fray. The pioneers of hookworm are relatively few in number and, happily, a rather congenial lot.

First up is Edorado Perroncito, an Italian veterinary pathologist, working in Turin. His findings emerged as a result of the construction of the Gotthard Rail Tunnel in Switzerland. This spectacular engineering feat connected Göschenen on the north with Airolo on the south. The nine-mile bore was begun in 1871 and completed in 1881. Approximately 200 men died during the decade of tunneling. Many were killed when their equipment hit veins of high-pressure hot water. The construction trains were another source of death. Of course trains burning oil or coal could not be used in a poorly-ventilated nine-mile tunnel, so drilling and travel power was provided by huge rolling tanks of compressed air.

This worked quite well, until a tank exploded. The third major cause of mortality (and much sub-lethal suffering) was severe anemia. Perroncito autopsied many of the dead and found them heavily infested with *Ancyclostoma duodenale*, a very close cousin of *Necator americanus*. He concluded, correctly, that the worms caused the anemia. The tunnels were ankle-deep in warm water; the workers had badly worn boots or none at all, and there was little in the way of latrines. All in all, the perfect breeding ground for an epidemic of worm infestation.

Next up in our pantheon is Arthur Looss (1861-1923), a celebrated German zoologist. He received his doctorate from the University of Leipzig, with a dissertation on trematodes (Parasitic flatworms, such as liver flukes). His pioneering work on bilharzia (schistosomiasis) in Egypt reduced illness in the construction of the Suez Canal. Farmers also suffered. (The annual Nile flood used to wash away the bilharzia but with the construction of the Aswan high dam more water is stagnant and farmers are, once again, infected.) Relevant to hookworms, Looss made two valuable observations, both of which are not for the squeamish. In his laboratory he had mixed up a container-full of growth media that no hookworm could resist—50:50 feces and soil. The third ingredient was hookworm eggs. In a few days, thousands of infectious larva bloomed. Each day they would rise to the surface and in huge bundles, began to move in unison, waving back and forth, seeking a host. (Looss termed this behavior "questing.") He accidentally touched his hand to one small portion of this witch's broth. Three hours later that part of his hand began to itch. Looss thought, "Could they have penetrated my intact skin?" Ever the scientist, he began collecting and examining his own feces. A few weeks later, there were hookworm eggs. In one moment of observation and thought, he had added two vital pieces of information about hookworms: they can penetrate intact skin and the period from infestation to production of the next generation of eggs is about three weeks.

Next up is Charles W. Stiles. Today, a person wishing advanced medical training has a whole spectrum of top-rank American institutions: Stanford, University of California, Harvard, and Johns Hopkins, to name just a few. But a century ago the situation was far different. In many areas of science, including biology and medicine, the shining light was in Europe, especially France and Germany. Stiles' training reflected that era. After a year of science (1885-1886) at Wesleyan University in Connecticut, he spent one year at the Collège de France, two years at the University of Berlin and another year at the University of Leipzig. He had further training at the Zoological Station of Trieste (then part of the Austro-Hungarian Empire) and at the Pasteur Institute in Paris. His Ph.D. was awarded at Leipzig. On his return to the United States he worked eleven years as a zoologist for the Bureau of Animal Industry of the Department of Agriculture. During those years he returned to Germany for twelve months (1898-1899) as medical attaché to the US Embassy. German swine producers had claimed that pork brought in from the United States was infecting Germans with trichinosis. Stiles' research satisfied the German authorities that US pork was safe. His credibility was no doubt enhanced by his fluent German and his Leipzig doctorate.

In 1902 he joined the Hygienic Laboratory of the US Public Health Service, where he spent twenty-nine years, co-authored a thirty-volume catalog of world-wide parasites, standardized nomenclature, studied Rocky Mountain Fever, analyzed health issues in cotton mill workers, and in 1923 was awarded the Public Welfare Medal from the National Academy of Sciences. He burst into the hookworm universe with his "Report upon the Prevalence and Geographic Distribution of Hookworm Disease (*Uncinariasis* or *Anchylostomriasis*) in the United States," published in 1903 by the Government Printing Office. (*Uncinariasis* was renamed *Necator*.) In this remarkable travel document he mentions neither his mode of transport, nor any tourist sites, but confined his words to the physical appearance of infected victims (thin arms and legs, swollen bellies, deathly pale skin, and dull-witted facial expressions) and the microscopic search for worm eggs in stool samples. He began north and traveled south. Summarized, he found the following.

District of Columbia. He and three assistants examined the stools of 500 male patients at the Government Hospital for the Insane (St. Elizabeth's). Fifteen patients were found to be infected with hookworms. Most had a connection with Cuba or the Philippines.

Virginia. With Dr. Charles V. Carrington, he examined 1,200 convicts at the State Prison and State Farm. They selected six white males who looked sick, but none had hookworm eggs. The 26 "Negro" inmates were also free of hookworm, although one was infected with eelworms. At the Richmond almshouse they also found no hookworm.

North Carolina. In the Virgilina Copper Mines, workers were not allowed to defecate in the mines. Most miners simply used the woods in a radius fifty meters from the shaft head. In stools collected from the woods, they found three with hookworm eggs. At the Cumnock Coal Mines, Chatham County, the miners were highly suspicious of these visitors. The one stool collected did show hookworm. At Sanford, Moore County, stools of four patients, collected by local physicians, showed no hookworm. At the Camden County brickyards, stools picked off the ground showed hookworm in two men, one white, one "Negro." At Haile Goldmine, Lancaster County, he obtained specimens from ten men. All were negative. At this point, the widely-read Dr. Stiles recalled data on Alaskan seal pups and outbreaks of worms in sheep and dogs raised on sandy soil. Local inquiry told him that there was a sandy district four miles from the mines. He examined a sandy-area eleven-member family; they all had the sickly appearance of infected victims. The stool sample examined was heavily infected with hookworm eggs.

At this point in his report, Stiles mused on the concept of "dirt eaters." Throughout the South, the constellation of signs and symptoms, described earlier as associated with hookworm disease, were ascribed to eating dirt. While there is little biological basis for this belief, it may have added to the comfort of the sufferers' social betters, that the dirt eaters had brought it on themselves.

Returning to North Carolina, Stiles visited the largest plantation in the sandy district near Kershaw and examined a sample of twenty workers (men, women, and children) laboring in the fields. All fit the description of "dirt eaters." He examined the feces of four of them and found hundreds of hookworm eggs. The plantation owner told the doctor that it would pointless to examine any further because all of his employees were "in exactly the same condition." Driving back to Kershaw, he stopped at a country school. About forty percent had the same general appearance as the plantation "dirt eaters." Back at Kershaw, Stiles met a farmer on the street who had come in from the country with two of his children. The record does not reveal how this Yankee stranger persuaded them to provide stool samples, but they did. All samples were "heavily infected" with hookworm eggs. The farmer assured Stiles that his entire family back home, ten in all, were just as sick. Stiles concluded that in Kershaw's sandy region the sickly appearances and the anemia (he does not state how anemia was measured) were all due to hookworm, confirmed by finding eggs in every sample examined.

South Carolina. At the Charleston Medical College, he met with the students and faculty (Stiles added "all white, of course") and explained his research. One faculty member and sixteen students immediately volunteered to participate. Four had hookworm eggs, while a fifth man was heavily infected with *Hymenolepsis nana* (tapeworm). All the hookworm cases came from sandy districts: Barnwell County; Saint John County; and Edisto Island. Stiles' next stop was the Charleston Orphan Asylum. "Through the courtesy of Dr. Huger and the ladies in charge, I was able to examine 230 white children, both boys and girls." Twenty were picked for further examination based on anemia and/or stunted growth. Fifteen of the twenty had hookworm. All of them came from sandy districts. (Here Stiles paused to thank the many South Carolina physicians who helped his quest.)

Georgia. At Atlanta he learned that local physicians, Claude A. Smith and H. F. Harris were studying hookworms at their own expense, so he made no studies there. "Furthermore, the territory belonged to them as local men." Dr. Lamartine G. Hardman, a member of the Georgia legislature, told Stiles that Jackson County had large numbers of citizens of "dirt eater" appearance. At Macon, Bibb County, through the kindness of two local physicians, he was able to examine the inmates of two white orphan asylums. In one, seventeen boys and girls, out of a total population of 85 children were selected for stool testing because of being "pale, weak, or otherwise poor condition." Of the seventeen, twelve children had hookworm and two had tapeworm. The hookworm cases were all from sandy districts: Americus, Sumter County; Buena Vista, Marion County; Thomasville, Thomas County; Savannah, Chatham County; Waycross, Ware County; Wacissa, Jefferson County; DeLand, Volusia County; and Liveoak, Suwanee County. At the second Macon orphanage, Stiles examined 112 white children; twenty-one looked infected. Stool examination of the latter showed seventeen cases were clearly infected with hookworm, one with tapeworm, and one with malaria.

The seventeen infected cases came from: Baxley, Appling County; Cordela, Dooly County; Darien, McIntosh County; Effingham County; Johnson County; Jones County;

Monroe County; Richwood, Dooly County; Sandersville, Washington County; and Waycross, Ware County.

Stiles moved on to four cotton mills near Macon, inspecting the mills and workers' houses, and found "25 or 30" cases of hookworm. Most had come from rural sandy districts. The longer they lived near Macon, the more their health improved. His next Macon visit was to a "Negro school," where he found no hookworm. On Macon's "Circus Day" Stiles stood on the sidewalk, observing thousands of pedestrians, many of them visiting from the surrounding country-side. Several whites looked infected, but none of the "negroes." Once again, Stiles praised local physicians for their cooperation.

At Milledgeville, he visited the "state Sanitarium" (opened in 1842 as the Georgia State Lunatic, Idiot, and Epileptic Asylum, remaining in operation until 2011) and with the assistance of Drs. A. M. Burt and L. M. Perry, saw a "large number of patients." Only two looked infested. Later stool samples showed hookworm in each. Stiles did not record the racial distribution of the inmates. At Fort Valley, Houston County, with the cooperation of Dr. M. S. Brown, he found ten cases of hookworm within an hour. The physical appearance was so typical, he examined only one stool specimen, which was heavily infested. Dr. Brown assured him that he could find fifty or more such cases within a day.

The next visit was to Albany, in Dougherty County, an area with heavy clay soil. The only hookworm cases were those in town from sandy districts. Dr. Hilsman, a local

Georgia
State Lunatic Asylum
DEC. 15, 1842

physician, noting Stiles' interest, drove him nine miles into a sandy region in Lee County. There they saw one family with four extreme cases. Microscopic analysis seemed unnecessary but was done for thoroughness. All four showed a plethora of hookworm eggs. While in Albany, Stiles made an error in diagnosis, which proved most helpful. The boy had malaria (typical history, enlarged spleen) but also appeared to have a medium-severe case of hookworm. However, microscopic examination showed no hookworm. After that Stiles made no more hookworm diagnoses based upon medium or light symptoms. Dr. Hilsman, from long experience, stated with "utmost positiveness" that there were no cases of hookworm in natives of clay soil areas. In his final Albany observation, Stiles stood on a street corner on Saturday, when the country folk from miles around came to town. In a two-hour period he observed roughly 200 whites and 3,000 "negroes." None of the latter looked infected, but five whites showed severe signs.

His next stop was at the southern Georgia county of Coffee, the local town being Willacoochee. The area, with sandy soil and numerous swamps, was a "center for dirt eaters."

There, with the help of Dr. Wilcox, he examined eight patients. They all looked like hookworm sufferers and all had stool full of eggs. Wilcox "declared that he knew of at least 200 similar patients within the territory of his practice." Stiles' final stop in Georgia was at Waycross, Ware County. Further north hookworm sufferers were designated "dirt eaters;" at Waycross they were termed "Branch-water people." Hookworm seemed "exceedingly common," and the local medical people "stated that they could easily show me scores of cases within a radius of a few miles."

Florida. At Jacksonville, Duval County, both state and city health officers assured Stiles that the condition he was studying was common throughout the state, and worse in the "flat woods district." At Waldo, Alachua County, Dr. J. W. Boring showed him two groups of typical sufferers within a short distance from town and described the condition as "exceedingly common." The investigation came to an end at Ocala, Marion County. There he easily found more cases through the assistance of Dr. A. L. Izlar. Five were confirmed by microscopic examination. All the Florida physicians ascribed the cause of this illness pattern to "malaria and improper diet." At Ocala, Stiles boarded a train and returned directly to Washington, DC.

Summary of Stiles' Findings. Hookworm was most prevalent in sandy districts, less so or absent in heavy clay soils. The further south Stiles went, the more hookworm he found. (Further north, frost kills the eggs in the soil.) Hookworm was extremely rare in African-Americans. Physical appearance could lead to a diagnosis, but only finding eggs in the stool could give a truly accurate diagnosis. Local physicians recognized the constellation of symptoms, but were not clear about the mechanism of hookworm infestation that crippled those infected. In his conclusions, he emphasized that he travelled as a zoologist, not as a physician, and left detailed symptom description, physical findings, and treatment considerations to medical doctors. His intention was to provide epidemiologic information to physicians, especially those in rural areas. The report, printed in pamphlet form, was widely distributed. However, Stiles' conclusions were, at first, mocked and ignored, as will be seen soon.

This report was very influential, after a frustrating delay of six years, in the 1909 formation of the Rockefeller Sanitary Commission, whose work will be described in a later chapter.

Our final hookworm hero was a son of the South, a native of Virginia's Shenandoah Valley, Dr. Charles Franklin Strosnider, known as "C.F.". His ancestors had come from Germany. They settled in the Cedar Creek Valley and worshipped at the Lutheran church in Woodstock. Many fought in the American Revolution. His great-grandfather moved from Cedar Creek to Strasburg. His parents married in 1878 near Orkney Springs in the Valley. Their first two sons died in infancy. C.F. was born December 16, 1881 and survived. He was still alert and active until he passed away in 1969.

The Strosniders had little reason to identify with Northerners. The forces of Maj. Gen. Philip Sheridan came through the Shenandoah Valley and took the Strosnider's serviceable horses, killed the foals, burned the barn, slaughtered the calves, and stole the household furnishings and silverware. His father, John Strosnider, served in the 10th Virginia Infantry. His uncles served as well: Henry and James in the 33rd Virginia (Stonewall Brigade); and Simon in the 39th Virginia Cavalry Battalion. However, the Civil War was not the first thing on C.F.'s young mind. He wanted an education and he wanted to be a doctor. In this, he was not encouraged. His parents wanted him to be a preacher and the family doctor warned of the expense and stress of medical school. C.F.'s early years of schooling were at the Strasburg Academy, from which he graduated in the ninth grade. As of 1967 he still had his McGuffey Reader. In the fall of 1900 he continued his education at the Oranda Institute, located in nearby Walnut Springs, which offered "Classical, Latin, Scientific, Normal, Business, and Bible Courses." He paid for his year at Oranda by selling a horse and six sheep that he owned. He next attended Massey Business College in Richmond. This young man planned ahead, deciding that with a few years in the business world he could save up enough money to pay for

1748

Strosnider, John.

Co. *A*, 10 Virginia Inf'y.

(CONFEDERATE.)

Private Private

CARD NUMBERS.

1	112042 16	18	
2	4281	19	
3	4344	20	
4	4407	21	
5	6670	22	
6		23	
7		24	
8		25	
9		26	
10		27	
11		28	
12		29	
13		30	
14		31	
15		32	
16		33	
17	•	34	

Number of medical cards herein. *0*

Number of personal papers herein. *0*

Book Mark:

See also

medical school. His local reputation for thrift and honesty enabled him to borrow $100 (a considerable sum then) from the Massanutten National Bank <u>on his signature</u>. At Massey's he took bookkeeping, shorthand, and typing. Records still extant praised his skill and diligence. He covered his room and board with two night jobs. After graduation he worked for a real estate firm for five months, earning a raise almost as soon they saw his productivity. He was then offered a position as a stenographer, in Wilmington, North Carolina, in the main office of the Atlantic Coast Line Railroad. There he was quickly promoted twice, with substantial raises each time.

Strosnider permitted himself no luxuries; every possible dollar went into his medical fund. He was then offered a lucrative position in real estate. Over the vociferous objections of his father, he turned down this chance and, on September 1, 1905 he entered the University of Maryland Medical School, at Baltimore. He did well. His junior year report card showed his lowest grade as 90, most of the rest were 99 or 100. His first two years cost $320 each. His outstanding studies earned him a full scholarship for his final two years.

Each student was required to deliver thirty babies before graduation. His junior year summer job was as an obstetrical intern at Walker Memorial Hospital in Wilmington, where he delivered forty babies. In that era most births were at home. In poor areas husbands threatened to kill the intern if either the mother or the baby died. Strosnider's obstetrical kit always included a club, for self-defense. (My friend and co-author, the late Jack D. Welsh, M.D. had similar experiences two generations later.) Strosnider received his M.D. in May 1909, second in his class, and passed the Maryland state board examination four weeks later. In July 1910 he was also licensed in North Carolina. The week following his 1909 graduation he went to work for the Rockefeller Sanitary Commission, formed to study and eliminate hookworm disease.

Although he was both a Southerner and a diligent, highly trained physician, this new role would not always be easy. North Carolina Governor Robert B. Glenn was the keynote speaker at the 1908 graduation ceremonies of Wilmington High School. He told the assembled youth, "A bunch of

Yankees are trying to intimidate us Southerners into thinking that we have hookworms. I'm telling you there is no such thing." Governor Glenn's image reveals strong conviction, but perhaps a tendency to counter-factual thought.

Doctors Perroncito, Looss, Stiles, and Strosnider raised the curtain on one of America's most successful public health programs. Politicians did little to illuminate that path.

CHAPTER 3
BILLIONAIRES AND WHITE TRASH

Dr. Strosnider worked with the Rockefeller Sanitary Commission for three years, 1910-1913. His many dramatic medical adventures are a vital part of the hookworm story, but they must be seen in the overall history of that organization. We will shall return to his story, but not just yet.

The many tangled threads of the Commission's history present several starting points. A good one might the story of Jekyll Island, Georgia. In 1888, a group of wealthy men purchased the entire island and turned it into a millionaires' playground. Many built "cottages," actually luxurious mansions., including Rockefeller's cottage.

At one point the combined holding of the members of the Jekyll Island Club were one-sixth of the wealth of the entire world. Rockefeller's story is well-known. Others are of relevance, both for their extreme wealth, but also for roles in the hookworm story.

Richard T. Crane (1832-1912) had his name on millions of toilets and bathtubs over several generations. He and his brother began manufacturing plumbing supplies and brass goods in 1855. Soon they were suppling steam heating equipment for the Cook County courthouse and the newly built Joliet Prison. By 1870, they had added elevators, engines, and steam pumps to their product line. In 1872 they were employing around 700 men and boys. By 1890 the Crane Co. was supplying the enameled cast-iron products found in millions of bathrooms across America. In 1910 the Chicago plants had 5,000 employees. In 1920, the company was the world's leader in the manufacture of valves. In the 1960s, they expanded further into aerospace equipment. The plumbing division was sold to American Standard Brands in 1990. In later years, Crane sported a white beard.

Another Jekyll Island notable was the financier Jay Gould (1836-1892). After brief schooling, he went to work in his father's store. Always ambitious, at night he taught himself surveying and mathematics. At age sixteen he opened his own surveying business. Over the next five years he helped prepare maps of New York's southern counties. After a few years in the leather industry he moved to Wall Street where he soon became successful, not just as a stockbroker but as a speculator. Operating though banks he was associated with, and by bribing judges and legislators he rapidly became a terror on Wall Street. In a vicious fight with Cornelius Vanderbilt for control of the Erie Railroad, Gould issued 100,000 shares of fake Erie Railroad stock, then made it "legal" by bribing New York state legislators. He then bought the Wabash Railroad, which carried wheat to New York's docks. To increase sales to Europe, by weakening the US dollar, he began to corner the gold market. Outfoxed by the US Treasury Department, Gould lost a fortune, but three years later was rich again. Buying railroads cheaply, and selling at a great profit, he made his way across the Midwest. While his railroads were pouring money into his coffers, he gained control of Manhattan's rapid transit system by buying the Elevated Railroad, and in a third masterstroke, seized control of the wildly profitable Western Union telegraph company. Biographers may differ on whether he was a brilliant businessman or a heartless robber baron, but he was clearly suited to buy a share of the Jekyll Island Club.

J. P. Morgan (1837-1913) was another larger-than-life figure. His admirers saw him as someone who had personally saved the country from two fiscal disasters, in addition to making industry far better organized. His detractors called him a robber baron who held the country hostage. Though he is gone, his shadow lives on: in the fourth quarter of 2015 JPMorganChase had a net income of $5.4 billion. He came from money and his education reflected that. From the Hartford Public School he went to the Cheshire Academy and then to the English High School of Boston, which specialized in math for future businessmen. After that, his father sent him to Bellerive in Switzerland to become fluent in French, and finally to the University of Göttingen, where he perfected his German and received a degree in art history. He entered banking in 1857 at the age of twenty. From then on he worked with or headed a large number of banking and investment firms. By the early 1900s he controlled fourteen large corporations, including US Steel, and twenty-five railroads.

He twice rescued the country from fiscal collapse. In the Panic of 1893 he sold gold to the Treasury which stabilized the Federal government. In the Panic of 1907 major New York banks were nearly broke. Morgan called together the major bank managers and the Secretary of the Treasury and forced an agreement in which Federal funds were transferred to the banks.

During the Civil War, he paid a substitute $300 to serve in his place. His other contribution to the Union cause was the purchase of 5,000 defective rifles, which he sold to the army at a 300% mark-up.

His vast wealth did not spare him two mortal sufferings. Childhood rheumatic fever gave him a lifetime of pain. More visible to the public was his rosacea which progressed to rhinophyma, a condition in which his bright purple nose blossomed with pits, fissures, nodules, and lobulations. He avoided photographers and all his official portraits were modified to show a normal nose.

J. P. Morgan, with his vast wealth, wish for visual privacy, and desire of a milder winter climate, fit perfectly into the Jekyll Island Club.

Henry Flagler was most directly connected to the Deep South. Through death and re-marriage, he had a step-brother, Stephen V. Harkness, who also became a multimillionaire. Henry's early career included a grain business and a salt mine. The latter went bankrupt when the end of the Civil War reduced demand for salt. Flagler returned to the grain business to re-pay his debts. There he met John D. Rockefeller, who was a commission merchant for the Harkness Grain Company. Within two years, Rockefeller moved to the oil business and with a loan from Harkness, arranged by Flagler, set up a refinery. The technical wizard was chemist Samuel Andrews. Thus was born the Rockefeller, Andrews, and Flagler partnership. They quickly crushed opposition by giving 15% rebates (kickbacks) to their customers. Rockefeller himself credited Flagler as the business brains of their success. Money was pouring in as Standard Oil became a near monopoly. In the late 1870s

¹³⁷¹⁶ H. M. Flagler's Special, first Train crossing Long Key Viaduct, 2.7 mi. long, Long Key, Fla.

Flagler's first wife, Mary, was quite ill and her doctor suggested winter in Jacksonville, Florida. A climate change was not enough and Mary died in 1881. Two years later, he married Mary's nurse, Ida Shourds. They honeymooned in St. Augustine, which they found charming, but very backward in hotel and travel accommodations.

To remedy these short-comings in 1886 Flagler began construction of the 540-room Ponce de León Hotel, and began buying and consolidating local railroads, to make travel easier for potential guests. He remained connected to Standard Oil, but devoted most of his time to Florida developments. He built the Royal Poinciana Hotel in Palm Beach and established West Palm Beach. He had intended to have West Palm Beach as the terminus of his railroad, but a devastating freeze convinced him to look even further south. Thus was born Miami, then mostly a swamp, but soon enhanced by the Florida East Coast Railway, a water and power system, a dredged Biscayne Bay, and its own newspaper. His final move south was construction of a railroad all the way to Key West. Beyond that, only ocean and Cuba.

His domestic life had more than a few riffles. In 1896 Ida was declared insane and placed in an institution. He then took up with Mary Lily Kenan, on whom he lavished a million dollars in jewels, back when a million was big money. But there was Ida. Flagler, a man of action, bribed the Florida state legislature to pass a bill allowing him to divorce Ida. Ten days after the divorce he married Lily.

WINTER HOME OF JOHN D. ROCKEFELLER.   ORMOND BEACH, FLORIDA.

(Rockefeller, of course, like any self-respecting billionaire, would have more than one Southern winter home. For years he travelled to many spots in Florida, escaping the Northern winters. In 1918 he bought a permanent *pied a terre*, The Casements, a huge mansion in Ormond Beach, Florida, where his guests included Edward, Prince of Wales, Harvey Firestone, Will Rogers, Thomas Edison, and Henry Ford. The latter two, prime examples of hyper-industrious, Horatio Alger mythology and Yankee industry, would have been especially filled with disdain for the lazy Southerners, the "poor white trash," the "dirt eaters.")

What, one might ask, was the connection of hookworm and Jekyll Island? Both Rockefeller and Flagler travelled through the South, usually in their private rail cars. True, they were isolated from the poor, but rail cars have windows and everywhere they went, in the slow-moving trains of that era, they witnessed at depots and in shanties along the right-of-way, the "dirt eaters." The whites of the South looked lazy, unhealthy, and unmotivated. They stared at the passing trains with empty eyes. The titans of industry wanted more customers in the South, but poor and listless people cannot buy things, much less oil and kerosene. Thus were the early seeds of hookworm suppression born. But first these titans had to learn that "poor white" meant disease, not moral weakness.

CHAPTER 4
THE SPARK AND THE FIRE

What was the mysterious leap between disdain and showering millions upon these poor whites? Did Rockefeller pull his train's emergency brake, wait until the screaming steel wheels fell silent, leap from his palatial rail car, embrace a barefoot tobacco-spitting local, and exclaim, "Let's be pals! I'll save you from disease and degradation"? Hardly. There were crucial intermediate steps. And most of these involved Frederick T. Gates.

The son of a Baptist minister, and a Baptist minister himself, in 1888 Gates was appointed secretary of the newly formed American Baptist Education Society. The following

year he met John D. Rockefeller, Sr., who was also a devoted Baptist. They hit it off immediately and soon he was soon

Rockefeller's senior adviser, not just on business matters, but on many philanthropic matters. In the former capacity he was, in his employer's estimation, a greater businessman that even Henry Ford or Andrew Carnegie. Much of the business end was handled through the houses of Kuhn, Loeb & Co., and J. P. Morgan.

More relevant to our hookworm story were the Rockefeller philanthropies and Gates' enthusiasm for the Efficiency Movement. This creed held that any charitable giving should produce the maximum benefit with the smallest expenditure; in today's coarse vernacular, "More bang for the buck." This credo would apply to the hookworm problem, but not for another six years. The intermediate steps are crucial to the story.

In April 1901, a trainload of high-minded Yankee millionaires set off from Manhattan to inspect some of the "Negro colleges" they had been funding. They were spurred by the hope of mending the continuing distrust between the defeated Confederates and the triumphant Union, and imbued with the fervor of white Northern protestant reformers. In southern Virginia the train paused briefly to take aboard Henry St. George Tucker, then president of Washington and Lee University. As the train rolled ever southward, he offered a word of advice to the self-righteous men of the industrial North: "If it is your idea to educate the Negro you must have the white of the South with you ... you must lift up the 'poor white' and the Negro together if you would ever approach success." Meanwhile, the millionaires' train clattered on, heading for a part of the nation with appalling illiteracy. Country-wide only 5 percent of Americans could not read or write. Among white Southerners the figure was 12 percent; among Southern blacks the figure was a shocking 50 percent.

As the train returned north, the partially-enlightened Yankees concluded that primary and secondary school were probably more relevant that a few heavily-funded elite black colleges. This ferment led to John D. Rockefeller, Jr. convening a meeting of how to best spend a million dollar grant just offered by his father. This conclave hatched the Negro Education Board, quickly re-named the General Education Board (GEB), with Dr. Wallace Buttrick, a Baptist preacher turned professional philanthropist, as executive secretary. In prior years he had made a survey of the black mission schools in the South, yet his underlying beliefs were little different from the most ardent segregationist: "The Negro is an inferior race—the Anglo-Saxon is superior." Another member of the GEB concurred: "The Southern black will willingly fill the more menial positions ... this will permit the Southern white laborer to perform the more expert labor."
(An objective historian, equipped with hindsight, might note that the complex iron castings which form the dome of our nation's capitol were made by highly skilled slaves, while the average white Southern laborer, in 1901, was so ill as to be capable of almost nothing.)

The Yankee capitalist evangelists soon began to realize that private philanthropy would not change a complex interlocking system. The South needed publicly supported schools. But to have schools, you had to have taxes. To have taxes, you have to have a functioning economy, and that requires a healthy work force. It would seem that Dr. Stiles' 1903 hookworm tour of the Atlantic Southern states would provide the answer. It did provide the answer, but no one would listen. A few months after his return he reported his findings and recommendations to a medical conference in Washington, DC. He was greeted not with applause, but with derision. The New York Sun scoffed at his ideas. Stiles was bombarded with hookworm jokes. Men with medical degrees derided him because he was mere zoologist. Even the renowned William Osler, a professor at Johns Hopkins and a Fellow of the Royal Society, denied that hookworm existed in the United States. (The author still has his physician grandfather's copy of Osler's *Practise of Medicine*, 1892 edition. The index does not contain "hookworm.")

Stile's quixotic search for someone who could comprehend the hookworm issue took a sudden turn for the better in 1908, in one of those moments which prove the adage, "It's not what you know, it's who you know." Theodore Roosevelt, in one of his many impulsive actions, appointed a commission on "country life." Included on the roster was Stiles. At a Southern railroad station, he and Walter Hines Page, both noted a pale, near-comatose man shuffling along the platform. Stiles remarked to Page, "That man is no imbecile. Fifty cents of drugs would make him normal." This thought did not go unnoticed. Back home, Stiles, used to the stultifying bureaucracy of Federal service, suddenly found himself on the dizzying roller coaster of Rockefeller World. Page introduced him to Buttrick. The three men "talked hookworm almost all night." The next day, Stiles was summoned to New York City, where he presented his findings again, this time to Gates and Simon Flexner. Half way through Stiles' slide presentation they halted his talk and called in Starr Murphy, a lawyer, Standard Oil protégé, and member of the GEB. Stiles began his talk again. Clearly, Starr, Gates, and Flexner understood and approved Stiles' thesis.

Within days, the senior Rockefeller put a million dollars of his personal money into the newly-formed Rockefeller Sanitary Commission for the Eradication of Hookworm in the South. The last eight words were quickly dropped because of Southern sensibilities, and the Rockefeller name was down-played because of the legions of Rockefeller haters. The Sanitary Commission was up and running.

Today, if a president (or governor or mayor) has a tangled problem, he appoints a commission. (A really dire problem gets a Blue Ribbon Commission.) The appointment creates the illusion of action. Over the months the commission acquires members, then staffers, then deputy staffers, then office space and stationery with fancy letterheads. After much activity, the commission issues a White Paper, which is rarely read and even more rarely acted upon. In the case of the Rockefeller Sanitary Commission the exact opposite prevailed. The commission was all men who didn't need a job, not political hacks, unemployed after the last election. Rockefeller's advisers knew their business. And what a lineup!

Of course, it included Stiles, the authority on parasites, and John D. Rockefeller, Jr., who was evolving into an efficient administrator, trusted by his father. But they are only the beginning of an illustrious line-up. There was William H. Welch, whose father, grandfather, and four uncles were all physicians. He studied Greek and classics at Yale and belonged to Skull and Bones. In 1875 he received his M.D. from the Columbia University College of Physicians and Surgeons. After a year of study in Germany, he returned to Bellevue Medical College, where he opened a new laboratory. In 1884, he was one of the founders of the Johns Hopkins School of Medicine and became the head of the Department of Pathology, with sixteen M.D.s working under his supervision. Men trained by him were eagerly sought by medical schools across the country. One of his students was Walter Reed. In 1916 Welch established the first school of public health in the country, and later founded the Institute of the History of Medicine. He began the *American Journal of Epidemiology*, and over the years was president of nearly every medical association or organization in the country. His Rockefeller credentials were solid: from 1901 to 1933 he was on the Board of Scientific Directors at the Rockefeller Institute for Medical Research. Finally, during and after World War One he was a brigadier general in the Medical Corps.

Simon Flexner was a native of Louisville, Kentucky, where he worked as a pharmacist for eight years. He then returned to college, receiving his medical degree in 1889. After post-graduate work at Johns Hopkins, he became a professor of pathology at the University of Pennsylvania. From 1901 to 1935 he managed the Rockefeller Institute for Medical research which, among other things, addressed the problem of hookworm in Puerto Rico. In 1907 he delivered a paper at the University of Chicago, predicting heart and kidney transplants. Edgar Gardner Murphy, a native of Fort Smith, Arkansas, devoted his life to education in the South. This long career began as he received his degree from the University of the South. (This remarkable institution was organized in 1857. One of its co-founders was Bishop Leonidas Polk, later a Confederate general. Another co-founder was Bishop Otey who wrote that the new university, "would materially aid the South to resist and repel a fanatical domination which seeks to rule over us.") Murphy was ordained an Episcopal Priest, served the church for twelve years and devoted the last ten years of his life to leadership roles in four significant organizations: the Southern Education Board, the Conference for Education in the South, the Southern Society for Consideration of Race Problems and Conditions in the South, and the National Child Labor Committee.

Edwin A. Alderman was born just three weeks after Lee's surrender at Appomattox. He graduated from the University of North Carolina in 1882, taught school in Goldsboro, and was soon superintendent of that school district. He successfully lobbied for the establishment of the Normal and Industrial School for Women, where he then taught for two years. Then came seven years at the University of North Carolina, where he was a professor and then president. He moved to New Orleans, served two years as president of Tulane University, and in 1904 began a twenty-seven year career as president of the University of Virginia. During his tenure there, enrollment quadrupled, departments of geology, forestry, and finance were added, and the first endowment fund was set up. His goal, mostly met, was to bring Southern education into the modern age.

H. B. Frissell was the first president of Hampton University. This remarkable institution had its origin in the Grand Contraband Camp, established by Union Major General Ben Butler, as a gathering spot for escaped slaves. In 1861, the American Missionary Association hired Mary Smith Peake, a mulatto who had secretly (and illegally) been teaching African Americans to read and write. She began her classes under a large oak tree, which still stands on the grounds of the modern campus. It was formally organized into a college in 1868. An early financial backer was Brevet Brigadier General William Jackson Palmer, a Quaker whose pacifism was out-weighed by his hatred of slavery. His brilliant and creative energy won him the Medal of Honor. He later used his talents to create the city of Colorado Springs. Well-known graduates of Hampton University include Booker T. Washington, Alberta Williams King, mother of Martin Luther King, Jr., and a host of successful athletes.

David F. Houston was born in North Carolina, graduated from the University of South Carolina, and did graduate work at Harvard. He was dean of the faculty at the University of Texas in 1899 and served as president 1902-1905, after a two-year interim stay as president of Texas A&M. This was followed by five years as chancellor of Washington University in St. Louis, where he strengthened the medical programs. Under Woodrow Wilson he served as Secretary of Agriculture and Secretary of the Treasury. After public service he held major positions with AT&T, US Steel, and the Mutual Life Insurance company. His thoughts on the South are explored in his book *A Critical Study of Nullification in South Carolina.*

Philander Claxton was born in Tennessee and received both his bachelor's and master's degree from The University of Tennessee. He furthered his education at Johns Hopkins and at universities in Germany. Thus equipped, he was superintendent of schools for North Carolina for ten years, then professor of pedagogy and German at North Carolina Normal College, where he was instrumental in running the Practice and Observation School, dedicated to the actual day-to-day skills of running a classroom. He returned to his native state in 1902, where he spent nine years as professor of education at the University of Tennessee. He continued his distinguished career by service as United States Commissioner of Education, under three presidents. His remaining years included important posts in education in Alabama, Oklahoma, and back to Tennessee.

James Yadkin Joyner was another of the educational statesmen who, in the early 1900s, revolutionized North Carolina's education policies. Orphaned at age two, he was raised by his grandfather, Council Wooten, who was greatly responsible for James' character formation. At age nineteen he graduated from the University of North Carolina with a bachelor's degree in philosophy. One of his friends and classmate was Edwin A. Alderman. After a year of teaching Latin, he moved to being superintendent of Lenoir County's public schools. After a brief stint at the progressive Winston school he entered law school. Following only one year of study he was admitted to the bar. Three years as a lawyer taught him that his heart belonged to education, not to the law. The years brought more posts and responsibilities. In 1902 Governor Aycock (another college friend) appointed Joyner as state superintendent of public education. The voters kept Joyner in his new post through five elections. His crowning achievement came in 1918, when the citizens of North Carolina approved a constitutional amendment raising the duration of the required school year from four months to six months. It is a sad commentary on Southern education that this was regarded as revolutionary.

The final member of the commission is one of the few Americans to be honored with a memorial plaque in Westminster Abbey, Walter Hines Page. He had been educated at Trinity College (now Duke University) and Johns Hopkins. After several years editing newspapers in the mid-South, he resigned and took up traveling and reporting sociological conditions in the deep South for three northern newspapers. In 1882 he wrote a series of important articles on Mormonism. Over the next thirteen years, he was active in New York literary circles, associated with many newspapers, book publishers, and magazines. He brought great names, including Rudyard Kipling, to American readers. In 1913, Woodrow appointed him ambassador to the Court of St. James, where he served very successfully during the First World War, resigning only because of serious illness. Britain's deep appreciation for America's help in the war, support channeled by Page, is still reflected on the walls of Westminster.

This commission was not just the varsity team, it was the Super Bowl team, an assemblage of the finest minds not just of the United States, but of any nation. And it, intentionally, it had a Southern slant, something very useful in avoiding the stigma of a "Yankee invasion," a term beloved of many Southern politicians. But this commission was not to do the day-to-day work of studying hookworm prevalence, working with local counties, hiring physicians, setting up clinics and dispensaries, and working with county officials and local doctors. For that work, the commission selected another Southern man, Wickliffe Rose, a man of great drive and impeccable credentials. He had degrees from the University of Nashville, the University of Mississippi, and Harvard. In the years 1902-1928 he was affiliated with, or ran, at least seven major organization devoted to health and education. In spite of his academic credentials and lofty connections, he was not afraid to go into the field, where he would see and hear of the ravages of hookworm up close. His remarkable reports, seen in the following chapter, illuminate his compassion, his intellectual curiosity, and his ability to listen carefully to his informants.

Dr. Rose, just before his epochal journey through the South, where he visited dozens of hookworm clinics.

CHAPTER 5
FIELD REPORTS

———————————

Wickliffe Rose in his role as administrative secretary was no desk-bound paper pusher. He travelled throughout the South and sent typed reports to "Mr F T Gates, 25 Broadway, New York City." They were more like chatty personal letters, much different from the cold reports that seem mandatory now, with their overuse of the passive voice ("Tests were performed..."). The author summarizes them here in a mixture of direct quote and paraphrase, all based on the digitized records of the Rockefeller Archives.

Northern Neck, Virginia June 1911

"On Thursday June 15, I met Dr Fisher by appointment at Fredericksburg to make a journey with him through the Northern Neck in order to see the results of our work in this territory where the field work first opened about fourteen months ago. The 'Northern Neck' is the neck of land lying between the Rappahannock and Potomac rivers and includes the counties of King George, Westmoreland, Richmond, Northumberland, and Lancaster. This territory constitutes Sanitary District No 1 in Virginia and has been assigned to Dr A C Fisher. From Fredericksburg we went by boat down the Rappahannock River for about 80 miles to Sharps; from Sharps we drove six miles to Emmerton, Dr Fisher's home. From this point we made a number of excursions on Friday and Saturday into the surrounding territory. Dr Fisher is the oldest man in the service; a Scotchman by descent with the energy, the tenacity, the hard common sense that belong to the blood. He was born in the Northern Neck; has been a country doctor here for more than twenty-five years; knows every man, woman, and child, black and white, in four counties and commands the confidence and esteem of his people."

When Dr. Fisher joined with the Sanitary Commission a year earlier he began with a survey of the level of hookworm infection in Richmond County. He selected one school in each district, collected a stool specimen from each child, and sent the specimens to "the laboratory" (laboratory location not specified). Averaging all the schools, 83 percent of the children were infected.

The infections were not evenly distributed. Some areas were almost free of infection, while in others nearly every single child was full of worms. The badly infected areas showed striking degeneration in physical, intellectual, moral, social and economic qualities.

They visited an area a few miles northeast of Emmerton, which contained an isolated population of "Forkemites," so named for the areas chief geographic feature, the widespread forks of a tidewater creek. They looked like a distinct race, with extreme poverty, lack of energy, pale complexion, absent thrift, low mentality, lack of moral perception, and dense illiteracy. These marks date back many generations. Dr. Fisher took Rose to the Totus Key school. Conditions there were so dreadful that nearly every teacher had quit. When surveyed a year earlier, 38 of the 40 students were infected. At home were 45 siblings who were so sick they had never come to school. All were dull, pale, and listless. Both the students and the children at home were treated. Their teacher, Henry Thrift, was eager to describes the changes. His pupils were now active and alert, not only studying but taking joy in studying. Rose's report describes "a new light to the eye, a new spring to the step, and a new outlook on life." To prevent recurring infection the school had new sanitary privy.

The tour continued. At a nearby farm they encountered William King, out plowing his field. Dr. Fisher called out, "Hello Willie, where'd you get that smile?" The 28-year old replied that he felt fine, was working every day, and had recently married. A year earlier he had been confined to bed, on the road to death, after 27 years of being an illiterate, thriftless, shiftless, anemic loafer. One round of treatment had transformed him. Across the road from Willie was the farm of Richard Prescott. For decades he had not been able to do a full day's work and his elder son was almost totally useless. Mrs. Prescott had been anemic since birth. She had born six children, even though bedridden for years. Now, when Fisher and Rose showed up she was out hoeing her vegetable garden and her husband was busy building a new house and a new sanitary privy. The oldest boy, previously a near-cripple, was out plowing, and planned to attend school that autumn, the first family member to ever have formal education.

Near the Prescotts were the Sydnors, husband, wife, and six children, two of whom had already died of anemia. No member of the family had ever attended school. They were not only sick and poor, but deeply in debt for medical treatments which had never helped. Dr. Fisher's work now had the backing of Rockefeller money. Treatment brought health, and to the Sydnor's astonishment, there was no bill to pay. Mrs. Sydnor, addressing Rose, lifted her arms to the sky and exclaimed, "It is the greatest thing that ever come."

Next on their itinerary was the cross-roads village of Haynesville, where W. R. Davis kept a store. He had grown up in a healthier locale and was "a fine physical specimen with plenty of red blood and a keen native intelligence." However, his children, born in the area, were all sick. A year earlier, Dr. Fisher had cured them and now their father was the community's greatest promoter of the hookworm campaign. The farms, fences, and fields were still badly neglected; the community was still awakening from generations of anemic stupor, but already Davis was seeing a double benefit. His neighbors were returning to normal life and in their increased prosperity had the money to shop at his store. "And so," wrote Rose, "the story might run on indefinitely. Dr. Fisher can tell you of cases like these all day long and show you the people."

Rose concluded his report with comments on the moral effect of treatment. The badly infected communities had become more and more isolated. They were too poor, too sick, to travel. Their neighbors shunned and avoided them. Physical and mental vigor faded and in their place came "an almost complete abandonment of the ordinary decencies of life." He continued, "I should like to give the story of a family or two to illustrate the moral effects of the disease, but the details are better omitted." (Was he referring to alcoholism? nudity? incest?) Rose's report on the Northern Neck, concludes with an optimistic, almost Biblical flavor.

"The results which I witnessed here are not only gratifying, they are stirring. I predict that within five years the whole face of the country in those pockets of extreme infection will be changed and one will see here a new people and a new earth."

County Dispensaries in Alabama May 1911

"On May 17, 1911, I met Dr Dinsmore by appointment in Montgomery to go with him into southern Alabama to see the county dispensaries at work. We were accompanied on the journey by Dr Leathers, State Supervisor of the work in Mississippi. On the 18th we visited the dispensary at Florala, in Covington county, where Dr Perdue treated on that day 186 cases. Dr Perdue is a young man; a fine physical specimen with plenty of red blood, a keen eye, an open genial face and a big heart; he makes friends because he can't help it and goes at his work as if this one thing he was born to do. He organized and conducted the dispensaries in Covington county; and in 18 dispensary days treated 2,504 people."

Dr. Perdue's dispensary at Florala was in a room 20 by 20 feet, with a few chairs and three tables.—one for his microscope, one for pamphlets and printed forms, and one for an exhibit of bottles of worms brought in by his treated patients. He also had a large number of doses of thymol, sealed in labelled envelopes. People began to arrive early and from 9 AM to 4 PM there was a great crowd. People had come from as far away as De Funiak Springs, Florida, a distance of twenty miles. They stared at the exhibits. They had their stools examined for eggs. They listened to the tales of those who had been cured, and went home to spread the news to their neighbors. Working in this way in Covington and Butler counties, in 25 days Dr. Perdue had treated 3,528 persons.

When the Sanitary Commission wanted to begin work in a new county, a representative approached the Commissioner's Court of a recently treated county and procured a letter stating how much money that county had provided. The Rockefeller people wanted the local governments to have "skin in the game," i.e., a sign of local political ownership and responsibility. Next he got a letter from the medical society of the treated county, stating the benefits to the patients and how it had added to the income of the local physicians. (An appeal to self-interest rarely fails.) This is followed by a letter from the Sanitary Commission central office in Montgomery to the prospective county medical society announcing that a representative of the State Board of Health would be at their next regular meeting, to describe the program and to ask for their support.

With this preparation, a representative of the Commission, often a physician, would arrive in the new county, ask the local doctors for their support and cooperation, in writing, both individually and as a society. With this endorsement in hand, the representative would approach the Court of Commissioners for a small appropriation from county funds, for drugs and printing. (Again, the importance of local financial support.) With this guarantee in hand, the "field man" would visit all the local physicians individually, inspect schools, give lectures at the schools, give public lectures using a stereopticon, supply the newspapers with press releases, and visit with mayors and leading citizens of each town. With endorsement from every venue, both private and

---

TWO HOOKWORM SPECIALISTS SPENDING WEEK IN ATHENS

Members of the Staff of the Rocke-
feller Commission on Hookworm
in the South.

Drs. Jacobs and Dobbs, two of the members in Georgia of the Rockefeller Hookworm Commission, are spending this week in Athens, investigating conditions here at the University of Georgia, the State Normal School, and the public schools of the city. Dr. Jacobs is a brother of Dr. Joe Jacobs of Atlanta—a former citizen of this city.

Athens Banner, 12 January 1911, p. 2, col.2.
©Athens-Clarke County Heritage Room, 2012.

This small newspaper item is an example of the publicity by which the Rockefeller Commission created interest in the hookworm program. Courtesy of the Athens-Clarke County [Georgia] Heritage Room.

governmental, the field man would then select three or four convenient spots for establishing dispensaries, designate what days each would be open, then flood the area with informative circulars. On the appointed days, the doctor(s) would examine and treat all who arrived. In most areas, nearly everyone who wanted treatment had been reached in under two months. Up to June 30, 1911, the combined efforts of Drs. Orr, Perry, and Perdue had organized a total of thirty-three free dispensaries, spread over seven counties, and had treated 11,466 persons. The seven counties contributed a total of $1,135.

Secretary Rose concluded his Alabama report with these words: "These county dispensaries accomplish two important results: They cure a large number of people in a very short time; they teach the people by demonstration how to recognize the disease, how to cure it, and how to prevent reinfection. It is the most effective educational agency yet discovered in this work."

Marion County, Mississippi

On February 27, 1911, Wickliffe Rose filed his report from Columbia, Marion County's seat of government. He began with a detailed summary of the area's economy, geography, and population. Columbia had a population of 2,000; the total population of the county numbered 23,651 in the 1900 census. The soil is sandy loam and clay. The climate is so moist and mild that most children, and many adults, go barefoot year round. Much of the land is still pine forests. In the cleared areas farming methods are crude. Men plow around the stumps, with defective equipment and an insufficient force of farm animals. Oxen still outnumber mules. The principal products are corn and cotton, The boll weevil has destroyed the latter. The valuation of personal property is $38 per person. The timber lands are owned by corporations that have headquarters far north.

Dr. Cully, who has charge of the Sanitary Commission's work in this district, estimates that 90 percent of the population has hookworm. He had visited six schools (Hickman, Edna, Hub, East Columbia, Cab Simmons, and Rogers High School) and examined microscopically the stools of 138 students. Of that number, 126 (91 percent) had hookworm eggs. Rose visited the last-named three schools and saw only one healthy-looking child in that entire group. Not one of those schools had a privy of any kind. The students simply urinated and defecated on the school grounds; nearly all were barefoot.

1st GRADE OF 1914 — Shown are the students who attended Columbia Primary School In 1914 (names correspond with numbers). (1) John Pope, (2) Van Morgan, (3) Francis Austin, (4) Clare Griffith, (5) Betty Colbert, (6) Harris Burkett, (7) Julia Kalil, (8) Sonja Kalil, (9) Ann Westerfield, (10) Jeanette Lawrence, (11) Miles "Booty" Roper, (12) John Davis, (13) Kenton Chapman, (14) Tom Morgan, (15) Vesta Martin, (16) Herbert Pope, (17) Ruth Eaton, (18) unknown, (19) Della Massey, (20) Nellie Speights, (21) Claudine Conner, (22) Tommy Griffice, (23) Fay Langston, (24) Charles Thompson, (25) Emily Turner, (26) Ruth Bennett, (27) Robert Watts, (28) Rivers Brown.

Three years earlier, ninety-one percent of the students at this school had hookworm. Now from this faded and yellow newspaper clipping, the students look healthy and alert, such a difference from the dull-eyed pallor of the typical *Necator* victim. They all have shoes, though this might be for the special occasion of their class portrait.

Cully informed Rose that when he first came to Marion County, he despaired. The infection was so heavy, and the poverty and backwardness was so great, that the task would be overwhelming for the local physicians. They agreed and supported the idea of a free county dispensary. The county health officer offered a free house for a year; the Commercial Club furnished the house; and the club ladies decorated it. The county board of supervisors paid for the drugs. When the dispensary opened on December 15, 1910, he worked all day and had to turn away hundreds. It is now open every Saturday and treats about 300 persons a month. Dr. Cully now spends most of his time "in the field" with a trained nurse running the dispensary.

Rose visited one school. The teacher and all the students had been treated. The teacher described the student's improvement as "very marked," and several boys agreed that they were much better. One teacher's eyes filled with tears as she described the positive changes. Rose visited the nearby Simmons family. Their eleven-year-old son had been given up for dead. Now he can cut wood all day. One man told Rose, "This is the greatest thing Mr. Rockefeller has ever done."

Rose concluded, "The people are beginning to understand that their infection, their lack of vitality, their unsanitary conditions, their poverty are connected; that the condition has come down to them and they are passing it along to the next generation."

North Carolina: Columbus, Robeson, Sampson, and Halifax Counties

Wickliffe Rose's report is dated August 14, 1911. A Dr. Ferrell had been the advance man; after his meeting with the Boards of County Commissioners of all four counties, they each had voted $50 to support a hookworm eradication program. Rose met with Dr. Covington who ran the program in Halifax County. Thus far a lecture at the school in Weldon had been the only progress. Most of the Halifax inhabitants were indifferent to the dispensary idea, and many thought it just scheme to enrich doctors. The doctors themselves were little better, in fact declared there was no such thing as hookworm disease. Ferrell set off on the necessary foundation building: a personal visit to each physician; meetings with the County Board of Health and the County School Board; prepared text and illustrations for the front pages each of the four county newspapers; and flooded the county with handbills, announcing the location and open hours of the dispensaries. He purchased envelopes, had dosage instructions printed on them, and filled each with thymol.

The Weldon clinic opened on July 25, 1911 at 9 AM; ten people were already waiting. During that first day, over 200 were examined and 55 received medication. A few days later, at Roanoke Rapids, he examined 180 persons and treated 113. Rose described some typical cases. Ira Scheerin came in with his wife and their thirteen children; all were infected and all were treated. Their extreme poverty had been made worse from doctor's bills for useless treatment. Grover Lane, age thirteen, came in with his father and his brother. "Back home nearly everybody looks like us. They used to say there ain't nothing in it, this hookworm thing but my other brother got treated last week at the other dispensary. He got all his worms in a bottle and showed 'em around to our neighbors. It certainly give 'em a surprise."

From Weldon, Rose took the train to Wallace and then drove twelve miles to Harrell's Store in Sampson County, to Dr. Strosnider's dispensary at the school house. As Rose neared the school, buggies were parked along the road, because the thirty-three buggies around the school occupied all the close-in parking space. There were at least a hundred persons in and around the school—all ages and all colors. With the doctor were his microscopist, J. J. Mackey, and a temporary assistant. All worked to their full capacity with no break for luncheon. Strosnider's cases for that week: Holly Grove, 116; Harrell's Store, 136; Ingold, 116; Roseboro, 227; and Clinton, 332.

The next week Rose, accompanied by Dr. Stiles, visited Dr. Page's dispensary at Fairmont, in Robeson County, and assisted him in examining and treating 170 persons in one day. Page's microscopist, Arthur McKimmon, was busy every minute. Later, Rose wrote at great length of the "appalling incompetence" of many local doctors. They did not hesitate to charge for wrong diagnoses and wrong treatment. One man had been treated for "piles," when in fact he had amoebic dysentery, diagnosed microscopically by Dr. Stiles. Rose was happier when he saw the county carpenter building two sanitary privies at a school, "to serve as an object lesson."

The following day, Rose took the early train to Fair Bluff in Columbus County to see "the experiment of the tent hospital and dispensary combined." When the conductor took his ticket, the man remarked: "You are going to the right place today. They are holding a hookworm convention up there." Looking around at his other passengers, he continued: "The whole damn crowd looks like they got it." Rose arrived at 8 AM and there were 83 people already waiting. The "Surgeon General" (Rose does not mention if this is the Federal Surgeon General) had provided four tents, twenty-three cots, and "all necessary equipment." That day, 217 cases were treated. Senator Brown (details not given) urged that the tent dispensary be continued. "I believe that would result in a complete eradication of hookworm disease in this territory." Rose cited two cases of special interest.

A boy age sixteen weighed only fifty pounds. He was held in his mother's lap, too weak to move. "Pale as a corpse and thin beyond belief." Treatment brought "marked improvement." On the train that evening Rose met a young man who had been an "extreme anemic." Now, after treatment, he had gained 45 pounds in six months, "and stood before me, the ruddy picture of health."

Rose praised the educational value of the clinics. People wanted to peer through the microscopes to see the eggs and to view the worms in bottles. At home, they told their neighbors, thus spreading the knowledge. At the end of this report he appended a summary table.

In the period July 15, 1911 through August 5, 1911: B. F. Dr. Strosnider's clinic saw 1,682 patients; Dr. B. W. Page's clinic saw 1,352 patients; Dr. C. L. Pridgren's clinic saw 3,047 patients; and Dr. P. W. Covington's clinic saw 1,169 patients. All in twenty-one days. Quite an achievement.

Guilford County, North Carolina

Wickliffe Rose visited this county in early November 1911 to see the workings of the newly formed County Superintendent of Health position. Hookworm was almost entirely absent, probably because of climate. The City of High Point is at an altitude of 1,000 feet and in winter the temperature is often below freezing. Neither the eggs nor the terrestrial worms will tolerate such cold.

Throughout its work, the Sanitary Commission had encouraged the concept of a paid, fulltime county health officer and Guilford was the first in the entire South to take such action, hiring Dr. Ross (first name not in this report), giving him an annual salary of $2,500, and even allowing him to purchase an automobile, perhaps one like this 1908 Ford Model T.

In the two days that Rose spent with Dr. Ross he saw firsthand the multi-faceted work of this new position. As a start, the law required him to inspect every school in the county for sanitary condition and certify such in writing, train teachers to identify conditions that required medical attention, examine any child with a debilitating condition of the eyes, ears, nose and throat, and to examine the feces of every child suspected of hookworm disease. If a child was found defective, Dr. Ross was to work with the parents to obtain treatment.

In the summer of 1911 there was an outbreak of typhoid, with 93 reported cases. Dr. Ross examined fifty wells and found forty-five of them polluted. All but two families cooperated in making their wells safe; there was one death among the hold-outs. As the school year progressed, Dr. Ross had visited nineteen schools, with a total enrollment of 902 pupils. One out nine students were reported for further examination. Nearly all of these had severe problems with tonsils and adenoids. The teacher was present at each examination and became the conduit for conveying information to the parents. "The teachers seem to welcome the opportunity to give this service." (Today's teachers might not welcome another duty of the 1911 schoolmarms: collecting fecal samples.)

The county smallpox situation had been "bad." Vaccination was not required. Dr. Ross pushed through a regulation that after January 1, 1912 all pupils must have a certificate of vaccination. Two isolated communities formed "anti-vaccination leagues." Dr. Ross visited them and with lectures and lantern slides secured greater cooperation. He also worked closely with mortuaries to secure better cause-of-death statistics.

Rose came away from this visit even more convinced that full-time county health officers were essential.

Bell County, Kentucky

In October 1912, Rose spent several days visiting a fully operational county-wide program. This report is of particular value in describing the techniques for visualizing hookworm eggs. Persons who wish to be examined are given a small vial (dimensions not described) in which to place a stool specimen. At the laboratory, the technician pours off roughly half the specimen, fills the container with water and stirs it thoroughly with a match stick. He (or she) then pours all the contents into a gauze-lined funnel. The gauze strains out the larger undigested food fibers and the screened specimen empties into a five-inch long tube, which is stoppered at the lower end with a cork. When the tube is full, he puts a cork in the upper end, thus sealing the tube, and places it in a centrifuge. After an unspecified time, he removes the tube from the centrifuge, takes off the top cork, pours off the fluid, and removes the lower cork, to which the eggs are adherent; this cork is then passed to the microscopist, who creates a smear on a microscope slide and searches for eggs.

Rose, unobserved, timed Miss James. She examined a new slide every 90 seconds. Miss Sigmeir, in Edmonson County, examined over 270 slides in one day. Rose himself took part in the technical activities. "At Pineville on October 5 I examined with the microscope a number of specimens submitted by persons who had had one treatment two weeks before. Of 44 specimens examined on that day, 36 were negative [no eggs]."

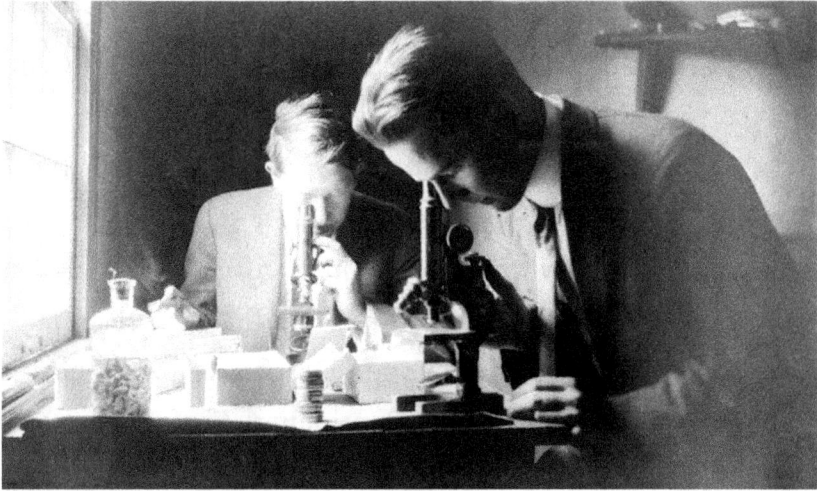

Elsewhere in his report, he described the remarkable community support for the hookworm project. A Mr. Frank Isaacs, who had been successfully treated, then brought in his family of seven, eight near relatives and twelve neighbors for treatment. He spent the day addressing the crowds, showing his bottle of worms ("All from my own family!") and exhibiting his son Roscoe, who had gained twenty-one pounds in four weeks after being treated. One of the most successful exhibits was one in which visitors could see live larvae wiggling under the microscope. A separate instrument was set us just for the visitors and a long line formed to see the display of dangerous parasites. A common reaction was, "I'll be durned." An old man who had hung on the fringes of the crowd all day, peered into the microscope just before closing time.

After a moment's reflection he approached Dr. Lock, pointed at the box of specimen containers, and broke his long silence. "Say, Doc. I want ten of them 'ar."

At Pineville, the Circuit Court was in session. Judge Davis suspended proceeding to have Dr. Lock address the crowd about hookworm. The county's forty-two physicians were very supportive of the entire program. Rose recorded a striking example of the economic benefits of public health. The general manager of the Continental Coal Corporation asked for help. At the company's camps there were 150 cases of typhoid and 65% percent of the men had hookworm. After appropriate sanitary measures, the coal loaded on the trains jumped from 600,00 tons a year to 800,000 tons, more than repaying the $25,000 the company had spent on treatment, sanitary wells, and latrines.

Summary

These letters to the Sanitary Commission vividly show both the extent of hookworm damage and the benefits of treatment and prevention. They record a historic moment when public health emerged as a vital part of government. They also help build the foundation for any assertion of the role of hookworm in the Civil War.

CHAPTER 6
DOCTOR STROSNIDER'S STORY

Chapter Two told us of his Southern heritage, strict rural upbringing, schooling, and medical training. As soon as he received his North Carolina medical license he went to work for the Rockefeller Sanitary Commission, commencing his hookworm years on July 10, 1910. He had been interning at Walker Memorial Hospital in Wilmington (see next page) when he heard of the newly organized hookworm program.

He quickly wrote to Dr. John A. Ferrell, assistant secretary to the State Board of Health, who had been appointed to head the North Carolina hookworm campaign. "I have been informed ... that you would possibly need some men to assist you in your line of work." Strosnider cited his experience in both medicine and surgery. "In reference to salary, I would expect over $100 and expenses, for the following reasons: 1st I am at present offered this amount with a corporation; 2nd If a man is on the road all the time, I have learned from experience that there are always incidentals arising which, in a way, are necessary for the success of his work; yet at the same time he does not feel justified in adding the same to his expense account."

James Walker Memorial Hospital, Wilmington N. C.

On June 16, Ferrell wrote back that he had submitted Strosnider's name to the Commission. A few days later, the request was approved. Ferrell wrote about the approval and invited Strosnider to a symposium on hookworm to be held at Wrightsville Beach on June 23rd. He further asked the new recruit to bring with him "three or four severe cases of hookworm disease that present marked clinical symptoms, make a microscopic examination verifying the diagnosis, and then endeavor to get these patients to go to Wrightsville Beach with one of us, at my expense." (In 1910 Wrightsville Beach had a permanent population of 22. Main activities were the fishing pier and the Carolina Yacht Club.) Apparently all went well, and Strosnider officially began work July 15, 1910, assigned to all of eastern North Carolina east of the ACL railroad. (Dr. Pridgen was assigned the Piedmont; Dr. Page was to cover the western portion of the state.)

Our Shenandoah native selected Goldsboro for his headquarters, taking a room at the home of Mrs. Tom Edmundson and getting his meals at Mrs. D. H. Dixon's, six blocks away. From the start, he had strong support from the editor of the Goldsboro *Argus*, Col. Joseph E. Robinson. Other publications were less helpful. The Raleigh *News & Observer* sniffed, "Many of us in the South are getting tired of being exploited by advertisements that exaggerate conditions." (Here we see the helpful Col. J. E. Robinson.)

The local medical profession was also unwelcoming. Some accused the Commission of trying to steal their private practice patients, while to others it was another power grab by Rockefeller, taking over the shoe industry and forcing Southerners to quit their time-honored custom of going barefoot. At age twenty-nine this young doctor, toughened by rural life and academic success through sheer hard work, was not afraid to beard the various medical lions in their dens.

He began in Fremont, Wayne County by introducing himself to the town's leading physician, Dr. L. O. Hayes. As soon as Strosnider explained his business, Hayes snarled, "My advice to you is to get out of town. The next train leaves at 3 o'clock and you better on it." Hayes huffed off to make a house call, leaving his caller standing there on the sidewalk.

By serendipitous coincidence, six Fremont college students, who had heard Dr. Stiles lecture on hookworm, recognized Strosnider as a "hookworm doctor", and asked to be tested. Good timing. Strosnider was carrying his microscope. He invited the students into Hayes' office and tested them. They were all positive. Just then Hayes returned and was furious. "What are you doing using my office? You have a lot of nerve!" His visitor handed Hayes the names of the six students. "They all have hookworm. They say you are their doctor. Treat them."

When it was obvious that Strosnider was not stealing his patients or questioning his authority, he calmed down and asked to meet that evening. Strosnider had chosen a boarding house because the landlady "always served ham and eggs for breakfast." They met there and talked until 10 PM. Hayes confessed that he did not know how to treat hookworm, and his visitor wrote out a prescription. The end result: the two men became good friends and the students got cured. The new hookworm doctor was clearly a master of what is now termed "community outreach."

There had been other barriers in Wayne County. The Sanitary Commission always asked for county financial support, in this case $300. The Wayne County Medical Society passed a resolution, asking the county commissioners not to vote any money for hookworm. The following week, the president of the medical society presented the resolution to the board. Adding, "I have a large practice and none of my patients have hookworm." When it was his turn to speak, Strosnider, who had been busy testing interested citizens, arose and read out the names of eight of the patients of the antifunding doctor. All of the eight had hookworm eggs in their feces. The medical opponents had no reply and the commissioners voted to help fund the work of the Sanitary Commission.

He faced even worse opposition in Jones County, where the chairman of the county board of commissioners was also a prominent physician. Thrice, Strosnider presented his request; thrice, he was rejected. On the third occasion, the chairman of the board told him not to return. The politician underestimated his foe.

At that time there was a ferocious battle for a seat in the US Senate, between the incumbent F. M. Simmons and

(In this photo, Simmons looks very senatorial.)

former Governor W. W. Kitchin. Strosnider approached the clerk of the county Superior Court and asked him if he would like to go about the county, campaigning for Simmons, with all expenses paid by Strosnider. In return, he asked to be introduced to two prominent citizens in the districts of each of the commissioners. By now, this determined young doctor had perfected his hookworm arguments. Every one of the "prominents" signed his petition. At the next meeting of the commissioners, Strosnider was allowed to speak, on the condition that he "keep it short." Short, indeed. He simply read the names of his new supporters. His opponents were outfoxed and outmaneuvered. By unanimous vote, the county paid their share. Unlike our current era, where politics are increasingly paralyzed by dogmatic partisanship, inability to compromise or forgive, and vicious personal attacks, in 1911, politicians could also be gentleman. The two opposing doctors went to lunch together, where the chairman conceded, "That was the slickest political trick I ever had pulled on me".

Strosnider was relentless in the campaign to slay the Great American Murderer, as he termed the disease. With A. T. Atkins, Wayne County Schools Superintendent, he visited every white school and every Negro school, giving lectures and leaving stool specimen containers. The teachers were to have students provide specimens, properly labelled, which were expressed to the State Laboratory of Hygiene at Raleigh. The teachers were notified of each infected child, information to be passed on to the parents.

One important finding was the racial distribution. Infections among white children were far more common than among the Negro children, and within the latter group, the darker the skin, the fewer the hookworms. Since the disease was brought to the Americas by slaves captured in Central Africa, a group quite resistant to the worms through millennia of evolution, this difference is not surprising. (The irony is hard to avoid. The light-skinned children had been weakened by "white blood.")

During his work in Wayne County, roughly 4,500 children were examined. A "high percentage" were found to be infected. In Grantham Township, out of 250 children at Falling Creek School, 248 were infected. Through lectures and work through the schools, Strosnider's campaign had a major impact on Wayne County. He moved next to Sampson County.

There, he established a clinic at Harrell's Store. His ability to turn rage into cooperation was shown when an irate farmer burst into the clinic, bellowing. "Where is that *** ****** doctor?" Having located the object of his interest, he bellowed again, "I'm going to beat you up. You said my daughter has hookworms!" The doctor replied that he was not afraid of this visitor and would fight him if necessary, but suggested first talking about hookworm eggs. "Our trained staff has found them in your daughter. When did she last attend school?" The answer: she had been too sick for two years to go to school. The outcome: the farmer, his wife, and all three children were tested, found positive, and successfully treated. The oldest daughter was so happy at her renewed health that she asked to visit the courthouse every Saturday to give testimony to her history and cure.

The doctor moved on to Pamlico County, where once again there was bitter opposition, including the smashing of the windows of the school building where he was holding a clinic. Once again, his personal skills turned opposition into cooperation. In a longer view, he was not just a "hookworm doctor." He viewed public health as a field with many problems. And solutions. He fought typhoid, both by vaccination and by cleaner water. He fought malaria with an anti-mosquito campaign. He spoke at clubs, teacher's meetings, churches, and schools He educated the public on tuberculosis, diphtheria, smallpox, and scarlet fever. He was a strong advocate of vaccination for smallpox and preventive inoculation for many other diseases, and raged against the anti-vaccinators, a counter-factual political force re-emerging in this century.

On January 31, 1913, he resigned from the Sanitary Commission to enter private practice at Mt. Olive, North Carolina. During his two years with the Sanitary Commission he treated an estimated 50,000 people who had hookworm. He left an astonishing and enduring legacy of public support for public health.

In the following fifty-seven years of medical practice, he continued to keep abreast of the changes in diagnosis and treatment, giving his patients the latest in medical care delivered in a very personal manner. He opposed the coming of Medicare, warning that it would entangle doctors in a web of regulations designed by nitpicking bureaucrats, removing the joy and dedication that marks real doctors. He could never have imagined that MDs would soon be HCPs (health care providers) lumped together with chiropractors, clinical social workers, respiratory therapists, and Christian Science readers.

He was no one-dimensional medical monk. He was active in every sphere of public life: teaching Sunday school, promoting agriculture and education, fulminating in medical politics. He enjoyed a "sociable" drink, and had several run-ins with Hell-spouting Methodist ministers. He loved playing poker all night, and was equally devoted to deep-sea fishing, duck hunting, and goose hunting. He survived typhoid fever, malaria, and an accidental arsenic poisoning. Two events saddened him deeply. His first wife, nee Rosa Meredith, died of uremia secondary to pregnancy. The baby died with her. His second wife, nee Nellie Edgerton, died in July 1926 of a pulmonary embolism while in labor. His third wife, nee Anna Lawrence, outlived him, dying at Duke Hospital in 1974 from metastatic bilateral carcinoma of the breast, with secondary osteoporosis and multiple fractures. Dr. Strosnider himself died of arteriosclerotic heart disease, with pulmonary congestion on December 9, 1969, and is buried in Riverview Cemetery, Strasburg, Virginia.

And how might all this be relevant to hookworm disease in the Civil War? Can these epidemiological figures be extrapolated back forty years? These questions will be considered in a moment. But not yet. Not yet.

CHAPTER 7
HOOKWORM AROUND THE WORLD

In 1911 the Sanitary Commission made a world-wide survey of hookworm infection, in order to find comparative data for their American studies. Other researchers have experienced difficulties in getting subjects to respond to surveys. The Rockefeller name was magic. The queries were done through the US Department of State, with strong support by the then Surgeon General. It became a diplomatic request, not just a bothersome query by some do-gooder agency. Result? Remarkable compliance.

(In many of the responses the provider used the term "coolie," which may be unfamiliar to younger readers. In the 1800s and early 1900s very large numbers of unskilled, illiterate workers from India and South China were recruited for hard labor in fields and mines all over the world. They were indentured workers, often recruited by deceptive means, and frequently mistreated in their new locations. Here are Indian coolie arriving in Trinidad.)

Western Hemisphere

Antigua: the local hospital in one year admitted 140 extreme cases of hookworm infection. All had been unable to work and 34 died.

Argentina: only 3 cases, all from other countries.

Barbados, West Indies: the infection is in agricultural workers; No numerical data is available.

Brazil: in the State of Sao Paulo, in 1909, there were 478 reported cases of hookworm death. Ninety percent of the counties reported deaths from hookworm.

British Guiana: fifty percent of the country was infected. The greatest problems were on the sugar plantations that used coolies from India. On one ship arriving from India, 78 percent of the workers were infected. One plantation owner treated all his workers. Production doubled.

British Honduras (now Belize): infection is general throughout the country. Of all cases of death in the hospital, autopsy showed hookworm in 70 percent. Most inhabitants go barefoot. There is no general scheme for detection or treatment.

Colombia: in the lowlands, 90 percent are infected. Agricultural workers and miners are the worst infected. Government officials take no interest either in the health effects or the economic damage.

Dominican Republic: there are no accurate figures but the disease is spreading. "The natives of the country are completely ignorant of the most elementary laws of hygiene." The government takes no interest in the problem.

Ecuador: Dr. Herman Parker, a US Public Health Service official noted that when he arrived in Ecuador he saw anemia everywhere. The cause was soon confirmed by hookworm eggs in the feces. Conditions are worst in the sandy areas. Sewage disposal is primitive. On one cocoa plantation only a third of the workforce was productive. The natives (not the Spanish descendants) seem free of hookworm. There is no government attention to the problem.

French Guiana: Dr. E. Brimont examined the inmates at five prisons. The average was 60 percent infection. "The disease has greatly retarded the development of French Guiana."

Guatemala: At the Ritalhuleu Hospital, 426 of 522 patients had hookworm. In 25 deaths, hookworms were the sole cause of death. On one coffee plantation over a two-year period, 258 had a confirmed diagnosis of hookworm.

Honduras: Hookworm was "prevalent in the interior and coast agricultural regions." There is no general public health information nor any governmental interest.

Jamaica: Total population was 820,000. Of this total, 16,000 were "East Indian coolies." Of these 50 percent were infected. A commission is studying the problem.

Martinique: *"Necator americanus* is prevalent throughout." Public meetings and literature are informing the public of disposal of night soil and other factors. A treatment program is underway.

Mexico: Hookworm is very prevalent in the southern lowland regions and rare in the drier northern states. Most Mexican doctors are incompetent and none seem to have microscopes. There is no government interest in hookworm. Mexican laborers imported into Arizona and New Mexico are a major cause of hookworm in the southwest United States.

Nicaragua: Infection is prevalent in the Bluefields region. There is no public health data and no public health service.

Panama: The US Consul General estimated that 20 percent of the population was infected. There is no government activity, either in collecting data or in initiating treatment.

Paraguay: The American consul wrote that hookworm was widespread. However, the country has still not recovered from the devastating five-year war instigated by the dictator Solano Lopez and such advancements as public health are utterly unknown.

Peru: Dr. J. C. Gutierrez, of the US Public Health hospital in Calao, reported that in forested regions thirty percent of the population has hookworm. There is no public effort to alleviate the disease.

Porto Rico (sic): In 1904, 80 percent of the population had *Necator americanus*. The infection rate in the coffee plantations was closer to 90 percent. Most Porto Ricans go barefoot their entire lives: the children go totally naked until age six. Since 1904, the Porto Rico Anemia Commission has treated hundreds of thousands of those hookworm victims and instituted major public health programs.

Salvador: Infection is heaviest among the coast Cordilleras and in the central portion. There are no accurate health statistics and no government interest in the problem. Most Salvadoran doctors are not acquainted with hookworm.

Surinam: The plantation laborers are mostly from India and Java; among them, the hookworm infection rate is 90 percent. Public health measures include installation of sanitary privies, a law against defecating on the roads and in the plantations, and "distribution of literature in the Hindostanese and Javanese tongues." Treatment protocol: A purgative in the evening; on the following morning four to six grains of thymol in pills, followed a few hours later by another purgative.

Trinidad: Our consul reported that the entire island was infected. There is no public health program. Cases are treated in the "estate hospitals."

Venezuela: Hookworm is probably widespread, but there are no reliable health statistics and the government has no interest in the problem.

Africa

Algeria: The agricultural oases are heavily infected. Workers go barefoot in warm moist soil. There are no health statistics, neither are there public health measures.

British East Africa and Zanzibar: The disease is known among the coastal natives as "safura." In Zanzibar alone, in a six-month period, there were 122 deaths from hookworm. The government gives treatment to "natives showing pallor."

Egypt: There are no reliable figures. At Kasr-el-Aing hospital in Cairo, 90 percent of autopsies showed hookworm. Army recruiters note a much lower infection rate among black recruits. Laborers work all day in damp warm soil; latrines are unknown. Health measures include treatment in government hospitals and improved latrines in mosques.

Gold Coast Colony: *Necator americanus* is more prevalent than *Ancyclostoma duodenale*. Infection is prevalent but there are no reliable figures. The government is encouraging improved latrines.

Natal (now KwaZulu-Natal): Hookworm eggs were first discovered here in 1906 by Dr. Boufa of Tougatt. The 1910 population of Natal was 700,000 natives, 80,000 Europeans, and 200,000 Indians. Recent ships bringing Indian coolies to work in the plantations had a 93 percent infection rate upon arrival. A quarantine has angered the planters. The Indians brought the disease and it is spreading west and north. "The Indian coolies are herded in barracks; they go bare-footed and wear scant clothing; their idea of sanitation and personal cleanliness is of the most elementary; it is a prevailing custom to keep wholesale scavengers attached to the barracks in the form of swine, hens and Muscovy ducks, the children, bare-footed and bare-bodied, play in the filth around the barracks, become heavily infected and cannot be made to conform to sanitary regulations." All infected immigrants are treated before being assigned to estates.

Sierra Leone: The only infection is among the poor of Freetown. There are no health statistics, nor efforts to control the infection.

Tunis: The oases of south Tunis are heavily infected but there are no exact numbers. "The dirt eating habit among these people is extreme; 60 percent of those found infected were confirmed dirt-eaters. The dirt-eater keeps a large store of his choice dirt at his house and carries a small bit with him wherever he goes." There is no government effort to study or treat the hookworm problem.

Asia

Bagdad, Turkish Province: The US vice-consul reports hookworm prevalent in agricultural areas. There is neither data nor efforts to control it.

Ceylon: The British authorities sent an eleven-page reply. Hookworm was brought by coolies from India. Nearly all are infected. They are reluctant to use latrines and merely squat where they are working in the fields. Hookworm is spreading to the Sinhalese. Government money should be used to treat all coolies. "Beta-napthol is the most suitable drug, thymol and other toxic anthelmintics be used only under medical supervision."

China: Nearly all the data is from the Medical Missionary Association. The southern two-thirds of China is heavily infected. Raw feces are the universal fertilizer. The government is of no help whatsoever. The problem is enormous.

Cochin China (Vietnam): No exact data but the Pasteur Institute in Saigon reports roughly 80 percent are infected. No government help.

India: The entire country is infected. An example: Surgeon Major Edwin Dobson examined 547 healthy-looking coolies from all over India; 454 were infected. Dr. C. M. Bentley in Assam examined 600 coolies; 599 were infected. Those who show up sick at government hospitals are treated, otherwise there is no general effort at treatment or eradication. There is also considerable infection with Kala Azar and Beriberi.

Japan: Surgeon Fairfax Irwin, US Public Health Service Hospital at Yokohama, submitted an elaborate report with tables, breaking down the data by prefecture and by occupation. Infection highest in the warm agricultural districts. Detected cases are treated with thymol. There is no systematic attempt at prevention or treatment.

Java: Reports from doctors all over the archipelago report very high levels of infection. One sufferer had a hemoglobin of 12 percent, barely enough to sustain life. Infections of all sort are abetted by the native custom of waste disposal —defecating directly into running streams and creeks.

Korea: Dr. O. R. Anison of the Severance Hospital in Seoul reports a high level of infection throughout the country. There is no government program for hookworm. The Korean Medical Missionary Association is studying the problem.

Malay States: The country has extensive rubber plantations, with Tamil, Javanese, and Chinese coolies. Nearly all have hookworm. The drains on the plantations are used for sewage, bathing, and drinking water. The government has sent letters to the plantation managers, urging sanitary measures. Dr. W. L. Braddon, State Surgeon at Serenbam, reported: "I am able to affirm that it is *to one single disease* (his emphasis) that almost all the mortality and sickness of the Tamil laborer is either directly or indirectly due. That disease is ancyclostomiasis."

Philippines Islands: There are no overall figures, but individual reports from all over the islands ranged from 12 percent to over 75 percent infected. In the rural areas there are no sanitary facilities, no privies. The householders simply defecate on the ground around the building; all are barefoot. There is no government program for hookworm detection or treatment.

Samoa: Assistant Surgeon P. S. Rossiter, USN, was instrumental in detecting hookworm disease on Tutuila and Manua Islands. His recommendations to the governor resulted in "every inhabitant of American Samoa ... supplied with sanitary facilities." The previous custom was to defecate on or along the road.

Straits Settlements (Singapore): In 4,000 autopsies, 10 percent showed hookworm. While infection was "prevalent," it is much less severe than in the Federated Malay States. Estate managers are encouraged to install sanitary privies.

Sumatra: Dr. J. Salm, colonial physician at Moeara, reports 96 percent infected.

Australia

Hookworm seems to confined to Queensland. Dr. T. F. McDonald found in one school along the Johnstone River, that 90 percent of the children were infected. In this district, hookworm is "sucking the heart's blood of the whole community." Dirt eating is prevalent. There are numerous cases of severe moral degeneration. The disease was brought in by "South Sea Islanders, Arabians, and Italians." Local authorities are given written material on prevention.

Europe

Austria: In 1903 there was a hookworm epidemic in the coal mines of northwest Bohemia. Strict government counter measures included rigid installation and use of leak-proof privies, requiring that the floors of the galleries be kept dry, and any infected person kept out of the mines until declared cured. By 1907, there was no more hookworm.

Belgium: There was infection in the coal mines and brickyards of the areas around Liége, Mons, and Charleroi. The problem was remedied by strict measures. Every new worker had to produce a recent certification that he was hookworm free. After one month of employment, he had a job-sponsored microscopic examination. Anyone found infected was treated until egg-free.

Bulgaria: As of 1911, no hookworm was detected. This is in part because any non-native worker must be microscopically examined before employment. Every year, every mine worker was required to have a microscopic stool examination.

France: Hookworm was confined to the mining districts. The original infection was brought by Belgian workers. There were immediate steps to control the disease. Every miner was examined before employment. Mine drainage and ventilation were improved. Movable waste buckets were provided near work areas. Sanitary privies were installed at the shaft entrances.

Germany: Hookworm was confined to miners. Between 1903 and 1911 infection was reduced 95 percent by the customary measures: pre-employment examinations; periodical re-examinations; and vigorous treatment of those found infected.

Italy: Major areas of infection are Sardinia and Sicily. Energetic measures have controlled the problem. The government printed and distributed free a pamphlet "written in simple language, and intelligible to the most limited intellect." Technical publications were sent to all relevant employers. All miners were periodically screened for hookworm. A "marine sanitary officer" screened immigrants landing from Brazil.

The Netherlands: Hookworm was found in the miners and brick workers in the Limburg region. Measures of examination and sanitation have reduced the percent of infection from 22 to 2.

Spain: Robert Frazer, Jr., American consul at Valencia reports heavy infestation in the Tabernes de Valldigna agricultural area. Local doctors see the problem as "anemia." In the mines of Linares, infection rate is 80 percent. There is no government effort for detection or treatment.

Switzerland: The lessons learned from the Saint Gothard epidemic were swiftly applied. In the digging of the Simplon and Lötscheberg tunnels there was no case of hookworm, nor is there anywhere else in Switzerland.

Wales: The only hookworm outbreak in Great Britain was in the tin mines of Cornwall. Treatment, education, and sanitary facilities brought almost total control.

California

To people on the East Coast, California might seem like a foreign country. The Golden State was certainly included in the Rockefeller publication *Hookworm Infection in Foreign Countries*. "Dr. Herbert Gunn, special inspector for the California State Board of Health, in his report on hookworm infections in the mines of that State, said, 'There is no question that the general efficiency of the men is noticeably impaired. At one mine, employing about 300 laborers, it was stated that a reserve of about 25 men had to be available to replace those who, on account of sickness did not appear for work. Quite a few of the men have to lay off every now and then to recuperate. Several who were unable to work stated that when they arrived in Jackson [county seat of Amador County] they were perfectly strong and well. A large number of these men were encountered on the streets, some of them presenting marked degrees of anemia. The greatest loss to mine operators is occasioned by the large number of those moderately infected. A loss of 20 percent in efficiency of those infected would be a conservative estimate. That would mean that in Mine No. 2, for instance, where over 300 men are employed at an average of about $2.50 per day, and estimating the number of those infected as low as 50 percent, a loss of over $20,000 a year.'"

Summing Up

It is manifest that in every warm moist part of the earth hookworm is endemic, with infections rates near 100 percent in some areas. When not controlled, it goes on generation after generation, always debilitating, often killing. Huge flows of population in the past few centuries have spread the worm to every continent except Antarctica. The great importation of African slaves to the New World and of Indian coolies to Ceylon and South Africa were major channels of spread. Natives of Sub-Saharan Africa have considerable resistance to the disease. Proper detection, sanitation, and treatment can yield total elimination, as in Switzerland. In 1911, the "developing nations," were making little or no effort towards that goal.

CHAPTER 8
HOOKWORM IN THE CIVIL WAR

The preceding chapters have shown hookworm infestation in every tropical climate, including the area covered by the Confederacy. There was little change in North American climatic conditions between 1865 and 1903. The South was still largely rural and the custom of going barefoot was still prevalent. (The author was surrounded by barefoot boys when he attended school in the South in 1944.) The soil that had been sandy loam was little changed in the intervening forty years. The only major differences were the mass troop movements and large concentrations of men. These, combined with the nearly absent modes of waste disposal, strongly suggest that in the war years, the rate of infection was even higher than in Dr. Stiles' 1903 survey.

And what about the presumably hookworm-free Confederate soldiers from dry clay areas and the Union soldiers operating in the South? Did they get hookworm? Probably. When the Confederate regiments assembled, the boys from sandy regions marched, camped, ate, and slept with the boys from the drier regions. The filth of the camps assured the mingling of dung, eggs, and feet. As for the Yankees as they entered the Deep South, anytime they were barefoot or shod in worn or wet shoes they, too, could become victims of hookworm.

Though not directly provable, there is a very high probability that hundreds of thousands of Confederate soldiers resembled the pale, anemic, listless men observed a century ago by Rockefeller and his friends. The Confederate officers, well-nourished, protected by boots, and riding high above the dung-filled roads and campgrounds, appear in ambrotypes and Daguerreotypes as pink-cheeked and healthy. Their loyal followers trudged after them, seized with the frequent bouts of dysentery that left their butternut trousers befouled with excreta, some dry and some yet moist. It could be said that this crusade for Southern "rights," was a vast army of poor whites led by dashing centaurs in gray.

The visual appearance of the common solders of the South was the inspiration for many unkind descriptions. The gaunt, skeletal "Georgia Cracker," beloved of Union cartoonists is one example. A Virginia woman, observing Robert E.

Lee's army headed north towards Gettysburg, remarked on "the gaunt starvation that looked from their cavernous eyes." William Faulkner's "Wash Jones," was described as "a gaunt ... man with pale, questioning eyes."

It would be unworthy to leave the Confederate soldier with just these descriptions. Indeed, he should be given additional credit for his ability and willingness to fight four years of war while feeling just awful, and General Lee should be further lauded for his charismatic role in inspiring these gaunt skeletons to follow him so far and for so long.

Away from the battlefield, hookworm played a part in the massacre of Andersonville. The lack of vegetables induced scurvy. The vastly inferior unbolted cornmeal was full of abrasive particles of cob and husk, which ripped and tore the lining of the gut. (Bolting removed the abrasive particles, leaving just the nutritious corn meal.) The water supply was filthy. And, finally, thousands of men, with thousands of loose bowels, were penned in a tightly-limited space, where they walked, ate, and slept in a swamp of sewage.

The <u>extremely</u> likely epidemic of hookworm is explored in deeply-researched detail in Dr. David F. Cross' "Why Did the Yankees Die at Andersonville?" (*North & South*, September 2003, pp. 26-32.) The Confederate surgeons sent to inspect the prison had many opinions regarding the high mortality rate but, knowing nothing of hookworm, they overlooked it. This iconic photograph shows a Union inmate of Belle Isle prison, but is representative of nearly every prisoner in Andersonville.

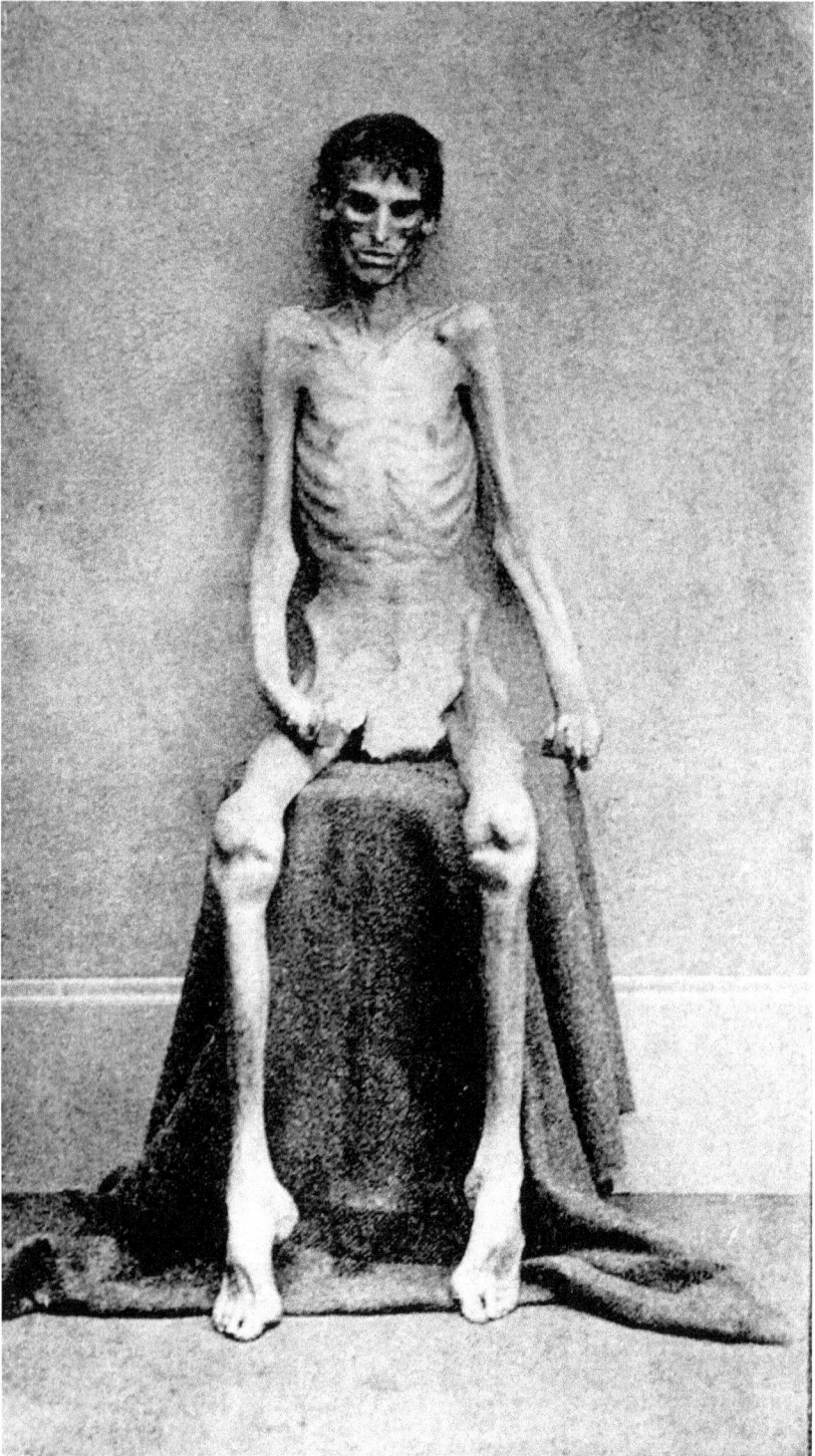

Leaping from the documented facts into the realm of speculation, and based on labor productivity reports all over the world, in which a minimum of 30 percent loss of productivity was widely reported, the author asserts that of the 250,000 men who served at one time or another in Robert E. Lee's Army of Northern Virginia, the equivalent of 75,000 men were lost to hookworm alone.

Returning to the primary question, did hookworm devastatingly weaken a very large portion of the Confederate soldiers? The answer is—yes!

APPENDIX A

ANEMIA AND THE WORM

In August 1951, just before returning to college, Randy Wilde, Sheldon Morris, and I traveled from the San Francisco Bay area to the foot of Mount Whitney in one day. We camped at the end of the road. Early the following morning we started up the trail and by noon we were at the summit, 14,505 feet above sea level. I did not enjoy the view. I was weak, dizzy, and on the verge of vomiting. Why?

At sea level the wonderful wisdom of the body (technically, "homeostasis") gave me just the right amount of oxygen-carrying blood to feel healthy. At the summit my blood was too "weak" to counteract the lack of oxygen in the thin air. By mountain standards, I was anemic. Anemia literally means "no blood," and was diagnosed, even though not understood, by noting pallor (paleness). Old time doctors looked at the nail beds and the tongue; they pulled down the lower eye lids and looked at the wet membranes. If these were a pale pink instead of a ruddy red the patient had anemia. Today's young doctors glance at the chart intently, at the patient briefly, and then hand him or her a lab slip. A technician draws blood, a complex computerized machine elsewhere does God-knows-what and spits out a report. Lest the doctor be too hurried to actually read the report, the computer helpfully marks the abnormal findings. That's how we diagnose anemia today. But enough of this *kvetching*. When and how did anemia become understood?

But first a note about blood itself. Take an ounce or so, put it in a centrifuge, whirl it, and soon it divides into two parts. The bottom half is a deep red; the upper half is a clear, yellowish liquid. The deep red is mostly the so-called red cells (corpuscles), mixed with a smaller amount of white cells, which exist mostly to fight infection. The red ones carry oxygen. The clear fluid is serum or plasma and has almost nothing to do with carrying oxygen. But how did we learn about all this?

It began with microscopes. And stains. The clear identification of red cells was noted by Dutch "natural historian," Antoj van Leeuwenhoek (1632-1723). He saw them but had no idea what they were. As microscopes developed there was one real problem: up close, body cells are almost transparent, much like clear gelatin. The colorful hues of Jello® are from added colors. The great advance in the microscopic under-standing of cells was parallel with red (or green) Jello®—the invention of stains. In the late 1800s, a series of brilliant men devised ways of making cells vividly visible. The interiors of stained white cells appear in shades of bright blue and pur-ple, and the red cells can be clearly seen to have a shape like a doughnut, with a thick rim and a thin (but not perforated) in-terior. With staining it now possible to count red blood cells. A modern lab report will show such things as: HCT (hemato-crit, AKA PCV, packed cell volume, the proportion of serum to the solid cells like red and white cells); MCV (mean corpus-cular volume, i.e., the dimensions of the average red cell); MCH (Mean corpuscular hemoglobin, the amount of hemo-globin in each red cell); and RBC (the total number of red cells in a given volume). There are several types of anemia which may be diagnosed by such blood counts, but this is not the place for a seminar in hematology.

What happened to the worm-bitten Southerners was a simple case of blood loss anemia. The hookworms were tak-ing blood faster than the victim could make more. The "poor white trash" felt like someone with altitude sickness—gen-eral malaise, weakness, trouble concentrating, and inability to "get organized."

When did doctors begin to understand what was hap-pening to these victims.? The 1854 *Webster's Unabridged Dic-tionary* does not even have the word "anemia." Nor does this dictionary mention "chlorosis," a term which has come and gone over the last century. Dorland's *American Illustrated*

Medical Dictionary (1951) tells us: "**Chlorosis:** Green sickness; a peculiar anemia mostly affecting girls about the age of puberty: so called from the greenish pallor of the skin. It is an anemia of defective blood formation characterized by a great decrease of hemoglobin accompanied by a slight decrease in number of red corpuscles. The disease is marked by perverted appetite, digestive impairment, debility, dysmenorrhea, amenorrhea, and nervous disturbance. **Egyptian chlorosis:** ancyclostomiasis (a type of hookworm)."

The introduction of staining led to an explosion of research and understanding. William Osler's *Practice of Medicine* (1892 edition) devotes twelve pages to anemia, with graphs illustrating the changes in red cells and hemoglobin in three different types of anemia: purpura haemorragica, chlorosis, and pernicious anemia. Osler discusses the two general categories of anemia (primary and secondary) and tells of great success in treating chlorosis with iron.

The essential points in hookworm and anemia are these. The hemoglobin in the red cells is the molecule which transports oxygen. The worm creates a steady loss of red cells, faster than the body can make more. The result is insufficient oxygen for the brain and all the rest of the body. The end result is a pale, weak person, unable to think clearly, move briskly, or feel any joy in life.

It is quite amazing that many of these pallid scarecrows marched all the way from Texas to Gettysburg, surviving on whatever they could glean from each successive neighborhood. They then fought, uphill, for three days in dreadful heat, against better-fed and better-shod Union men. The Confederates who were not killed then escaped south across the Potomac to fight another day. Whatever one may think about their politics and causes, as a group Lee's army illustrated the ability of men to push their exhausted oxygen-short bodies to the limit.

APPENDIX B
SHOES

Men with shoes, good shoes, were protected from hookworm. They were also, in theory, more able to endure long marches over rough roads and sharp stones. In actuality, Civil War era shoes were a source of misery. There were no left or right shoes, until later in the war, just ambi-shoes. There were often just two sizes: large and small. The chances of a comfortable fit were minimal. A modern commentator has remarked, "If one side had had New Balance, or Nike, or L. L. Bean shoes, the war would have been over in a year." Whatever the discomfort, both sides wanted shoes.

By the 1860s the hand-crafting of shoes, using technology many centuries old, was quickly giving way to shoes sewed by machine. In 1861, in the North, there were 74 firms manufacturing sewing machines. In the South, none. If the South was to have an army of shod soldiers, it would need an army of cobblers. The answer was to import shoes from England. The blockade by the Union Navy captured many ships but the majority got through. Documentation is difficult as the trade was illegal; the Southern war profiteers, the British exporters, and the Bahamian middlemen all wished to remain anonymous. However, the trade was vast as shown by the following figures: in a five-week period in late 1864, the South imported from the Bahamas 3.6 million pounds of meat, 1.5 million pounds of lead, 316,000 blankets, 542,000 pounds of coffee, 69,000 rifles, and 450,000 *pairs of shoes*. (Lowry, Thomas P.: "The Big Business of Bahamian Blockade Running," *Civil War Times* May 2007).

Assuming these figures are accurate that would be six pairs of shoes for every man in the Army of Northern Virginia. What was wrong? In a word, no Haupt. In April 1862, Lincoln commissioned Herman Haupt, a West Point graduate and experienced railway engineer, as a colonel and gave him full power in the running of the Union railroads. With an iron hand, he brought order out of chaos. The South had no Haupt; instead, they had States Rights. This principle guaranteed that the various governors and privately held railroads would never coordinate schedules or track sizes. There were shoes but they rarely got to the soldiers.

In the 1840-1860 era, what we today would call shoes were termed bootees, a designation now reserved for infant's

knitted foot coverings and, in some circles, a part of the ana-
tomy. During the Mexican War, there was such a problem
with crooked contractors that the Army set up its own factory
at the Schuykill Depot. In June of 1861 it had 700 employees
and a year later was producing 15,000 pairs of bootees a
month. Early in the war, Maj. Gen. Halleck reported from the
West that soldier's shoes wore out after only four days of
marching. To further complicate things, most shoemakers
were producing pegged shoes and were inexperienced with
sewing shoes. However, by the middle of the Civil War, with
the aid of Gordon McCray's patent shoe sewing machine,
which rapidly connected the soles with the uppers, production increased and Union men now had more reliable sources
of durable shoes. (Risch, Erna: *Quartermaster Support of the
Army*. Washington, DC. GPO, 1962.)

The movie *Gettysburg* shows the viewer many pairs of
bare Confederate feet, but how widespread was the shortage
of shoes in the armies of the South? There is, in fact, no general source of statistics regarding shoes in those armies.
There are, however, some statistics for certain units at certain times. An example, provided by Robert K. Krick, the leading historian of the Army of Northern Virginia, concerns the
brigade of General John Bratton. Inspection reports, still on
file as NARA microfilm M-935 report equipment and short-
ages of such in late 1864 and early 1865. Bratton's Brigade
contained five regiments: 1st, 5th, and 6th South Carolina Infantry, 2nd Rifles, and the Palmetto Sharpshooters. On September 8, 1864, the 1st South Carolina had 215 men present

for duty, 177 present for inspection, and 503 "Aggregate present and absent." The other four had similar figures. The inspection of November 26, 1864 showed a total of 1,226 guns in the brigade, all .58 rifled muskets. There were deficiencies of 286 coats, 500 "trowsers," 600 blankets, 370 haversacks, and 380 shoes (pairs of course). (In the following inspection reports, only shoes will be listed.) December 27, 1864, deficient 160 shoes. January 27, 1865, deficient 466 shoes. February 25, 1865, deficient 354 shoes. In brief, in several months, a third of the enlisted men were "deficient" in shoes. ("Deficient" probably includes both soldiers with deficient, worn shoes and soldiers with no shoes.)

The shortage of shoes, coupled with diarrhea, exhaustion, heat, and days without food greatly impaired Lee's army as they approached Sharpsburg (Antietam). (The effects of hookworm were not documented, as its existence was unknown.) Commentators on the march left these descriptions: "The ambulances were full, and the whole route was marked with a sick, lame, limping lot, that straggled to the farmhouses that lined the way," and "men exhausted by the rapid march and overcome by the dust and heat, fell out of the ranks and were left along the roadside by dozens." All told, Lee was missing over forty percent of his troops as he arrived at the battlefield. It is hard to imagine the utter misery of a Civil War forced march. (D. Scott Hartwig: "Lee's Shrinking Army." *America's Civil War* May 2016.)

From the documentation available, there was a shortage of shoes in the Confederate armies. The exact numbers over the course of four years of war is unknown.

A BIOLOGIST AT GETTYSBURG

Human beings are many things—moralists, philosophers, dreamers, lovers, haters. They are also mammals, subject to all the forces of genetics, upbringing, experiences, weather, injuries, and parasites, as are other organisms.

One of the classics of post-war Civil War literature is *Battles and Leaders*. It contains many essays by former officers and opinions and memories of many battles. In this brief meditation on hookworm, there is only one battle, and that is Gettysburg. The leaders to be considered from a biologist's point of view are Lee, Longstreet, Meade, and Hancock. Then will come the turn of the soldiers and in particular the 13,000 men who formed the immortal Pickett's Charge.

Robert E. Lee was age fifty-six. A year earlier he had had a bad cold and was very tired for at least a month. In August 1862 he had a bad fall, breaking bones in one hand and spraining the other. One hand was splinted and he was unable to ride his horse for two weeks. A month before the battle he had another bad cold and his doctors moved him from his tent to a house. He had pains in his chest, back, and arms. Modern reviewers doubt the diagnosis of pericarditis. By mid-April he had recovered enough to return to camp., but was still very weak and could do little work. During the actual battle he had diarrhea. Was he also plagued with lice? Almost certainly. In nice middle-class areas today children come home with head lice. Lice were widespread in both armies. They are no respecters of rank, courage, or charisma.

James Longstreet was age forty-two. He was wounded so severely in the Mexican War that he was unable to travel for two months. In September 1862 a boot chafed his heel so badly that the skin did not heal. He was forced to wear a slipper and ride sidesaddle at Sharpsburg. Four months before the battle at Gettysburg he was sick with a sore throat that confined him to his quarters. He was not wounded at Gettysburg but was wounded in the neck and shoulder in May 1864. Bloody foam poured from his mouth. Five months later he was able to return to duty, although with a paralyzed right arm.

George Gordon Meade was forty-eight. As a West Point cadet he had been treated for headache, sore throat, sore hand, a boil, a fever, rheumatism, nausea, influenza, and vomiting. In 1836 he had severe fevers while serving in Florida. In 1845 he was sick several months with jaundice. In June 1862 he was severely wounded at Frayser's Farm, Virginia. His arm healed fairly quickly but the severe back wounds damaged his liver and one kidney. He returned to duty six weeks later "much prostrated." He served at Second Bull Run, Antietam, and Fredericksburg, though still weak. Five months before Gettysburg he was off duty for three weeks with pneumonia and was not entirely well when he assumed command of the Army of the Potomac.

Winfield Scott Hancock was thirty-nine. In Mexico he received a contusion at Churubusco and was very sick with "chills and fever" (probably malaria) at Chapultepec. At Fredericksburg a musket ball grazed his abdomen. At Chancellorsville he was struck by several small shell fragments. At Gettysburg a minié ball passed through the pommel of his saddle and into his thigh. A surgeon used his finger to remove wood fragments and an iron nail from the wound. A year later, bone fragments were extruding from the unhealed wound. Years later, after many tries at returning to duty, it still troubled him. When the battle began, he seemed to be only member of this quartet who wasn't already sick.

These histories are from Dr. Jack D. Welsh's magisterial *Medical Histories of Confederate Generals* (Kent State University Press, 1995) and the equally essential *Medical Histories of Union Generals*, which followed in 1996. Today, most of these men would been given early retirement, as too sick or too crippled for active service.

Now the scene shifts to the line of trees where the soldiers of the Army of Northern Virginia are concealed. Those who know the coming orders glance across the mile of open field to that immortal "clump of trees," and weigh their fate. Thirteen thousand men had been chosen for the charge which will immortalize the name of George Edward Pickett (himself only partly recovered from a severe shoulder wound at Cold Harbor.)

From the post-war studies elaborated in earlier chapters, it likely that 90 percent of those men sheltered in the line of trees had hookworm, giving the reckoning that the soon-to-be-launched charge would contain 11,700 infected men. At 2,000 hookworms per soldier, the Yankee defenders were faced with an approaching wave of approximately 23 million hookworms. There are hundreds of books and thousands of articles about Gettysburg. It is possible that this aspect of the famous battle may have been overlooked by other writers.

APPENDIX D
KILLING THE WORM

In 1910, the doctors busy killing hookworms were using thymol, combined with Epsom's salts. A doctor today would use albendazole. Here are the details.

Thymol has an ancient history. It was used in pharaohonic Egypt to preserve mummies. In classic Greece it was burned in temples, with a scent pleasing to the gods. In medieval times it associated with courage, and women created a "pledge" for their courtly man with an embroidered bee and a sprig of thyme. Thymol is found in the common thyme plant, *Thymus vulgaris*, and in many other plants such as the bee balm species, *Monarda fistulosa* and *Monarda didyma*. The chemists were busy with thymol well before the hookworm doctors took an interest in it. In 1719 the German chemist Caspar Neumann isolated thymol from the other elements of the plant. Eight years before the fire eaters and the abolitionists set off America's worst disaster, French chemist A. Lallemand named this extract thymol and gave it its chemical name, 2-isopropyl-5-methyphenol. Bypassing Mother Nature, synthetic thymol was created by Swedish chemist Oskar Widman in 1882.

A common regimen of thymol in the removal of hookworms was a purgative/laxative such as Epsom's salts at bedtime, with thymol six to eight grains in the morning and a second dose of purgative in the late afternoon, to wash out the dead worms. (One grain equals 60 milligrams.)

Epsom's salts get their name from a bitter saline spring at Epsom in Surrey, England. Its chemical identity is magnesium sulfate. In addition to its uses as a purgative it is widely used in agriculture to stimulate plant growth.

An older generation will remember the term "Like a dose of salts," meaning something happening rapidly. Current readers who have had a colonoscopy will have vivid memories of magnesium sulfate's powers in emptying the human intestinal tract.

Hookworm today is usually treated with albendazole or mebenazole. A single dose of the former is efficacious in 72 percent of those treated, while the latter's efficacy is only 15 percent. It would seem that albendazole (trade name Albenza) is the preferred remedy.

SOURCES

Anonymous: *Hookworm Infection in Foreign Countries.*
Rockefeller Sanitary Commission, Washington, DC.
1911.

Chernow, Ron: *Titan: The Life of John D. Rockefeller, Sr.* Random
House, New York. 1998.

Cross, David F.: "Why Did the Yankees Die at Andersonville?"
North & South magazine. September 2003. Vol. 6,
Number 6, pp. 26-32.

Despommier, Dickson D.: *People, Parasites and Plowshares.*
Columbia University Press, New York. 2013.

Drisdelle, Rosemary: *Parasites—Tales of Humanity's Most
Unwelcome Guests.* University of California Press,
Berkeley. 2010.

Josephson, Matthew: *The Robber Barons*. Harcourt, New York. 1934.

Rountree, Moses: *Hookworm Doctor—The Life Story of Dr. C.F. Strosnider*. Nash Printing Company, Goldsboro, N.C. 1968.

Stiles, Charles Wardell: *Prevalence and Geographic Distribution of Hookworm Disease (Uncinariasis or Anchylostomiasis) in the United States*. Treasury Department Public Health and Marine Hospital Service. Hygienic Laboratory Bulletin No. 10. Government Printing Office. Washington, DC. 1903.

Stiles, Charles Wardell: *Country Schools and Rural Sanitation— Six Sample Public Schools in One County*. United State Public Health Service Reprint No. 118. Government Printing Office. Washington, DC. 1913.

Front Cover Photographs

These are scanning electron micrographs, provided by Dennis Kunkel Microscopy, Inc., of Kailua, Hawaii. The center image is *Necator americanus*, with its cutting plates in the upper portion of the mouth. The side images are of *Ancyclostoma duodenale*, a closely related hookworm, with its prominent "teeth." Images are copyright, Dennis Kunkel Microscopy, Inc.

INDEX

Idle
Winter
Press

www.ingramcontent.com/pod-product-compliance
Lightning Source LLC
LaVergne TN
LVHW021343080426
835508LV00020B/2095